Cláudia Cristina B. Ramos Naizer

Passenger transport demand analysis for the metro network

Cláudia Cristina B. Ramos Naizer

Passenger transport demand analysis for the metro network

Case study for the city of Salvador - BA

Imprint

Any brand names and product names mentioned in this book are subject to trademark, brand or patent protection and are trademarks or registered trademarks of their respective holders. The use of brand names, product names, common names, trade names, product descriptions etc. even without a particular marking in this work is in no way to be construed to mean that such names may be regarded as unrestricted in respect of trademark and brand protection legislation and could thus be used by anyone.

Cover image: www.ingimage.com

This book is a translation from the original published under ISBN 978-613-9-63681-5.

Publisher:
Sciencia Scripts
is a trademark of
Dodo Books Indian Ocean Ltd. and OmniScriptum S.R.L publishing group

120 High Road, East Finchley, London, N2 9ED, United Kingdom
Str. Armeneasca 28/1, office 1, Chisinau MD-2012, Republic of Moldova, Europe
Printed at: see last page
ISBN: 978-620-7-69214-9

DEDICATORY

To my family and friends

SUMMARY

The aim of this work is to analyse the demand for passenger transport on the Salvador - BA metro and to propose a metro network that is suited to the city's needs. It proposes a metro network that adequately serves the main centres of the city of Salvador and that drives the migration of demand from car users to the metro. The study was based on the sample of the Origin/Destination survey carried out in the city in 1995. The Four Stages model was used as a demand modelling tool, focusing on the first two, Trip Generation and Trip Distribution. In the Trip Generation stage, models were obtained to forecast future trips produced and attracted using the multiple linear regression technique, considering socio-economic variables as independent variables. In the Trip Distribution stage, the number of trips between each pair of traffic macro-areas was estimated, making up the future Origin/Destination matrices from the trips projected in the previous stage. The traditional Fratar and Gravitational models were used for Trip Distribution. The results of these stages were presented in the form of maps of desire lines, highlighting the origin-destination pairs with the highest demand. An analysis was made of whether the current network covers the regions in question and an alternative network was proposed, which covers the main journeys by public transport and car in the region. It was concluded that the current network is consistent with the main axes of expansion of the city of Salvador, but does not adequately serve some important sub-regions. Connections are proposed to serve a large number of private transport users. In addition, a connection is proposed between the suburban train and Line 1 of the existing Metro, in order to connect the suburbs more directly to the city centre. Although the study is limited to the municipality, it would be interesting for Line 2 of the Metro to be extended to the neighbouring city of Lauro de Freitas.

Keywords: Passenger transport demand analysis; Metro network; Salvador - BA

Summary

CHAPTER 1		**4**
CHAPTER 2		**7**
CHAPTER 3		**12**
CHAPTER 4		**28**
CHAPTER 5		**60**

CHAPTER 1

Introduction

1.1 Background and justification

From the 1950s onwards, the Brazilian population's commuting pattern changed due to the country's industrialisation process, which caused the large urban centres to grow rapidly. In a short period of time, Brazil ceased to be a rural country and became predominantly urban.

With urbanisation came its problems, such as the daily commute from home to work. The unequal organisation of the territory of today's metropolises creates a scenario of dispersion of origins, households, and a concentration of destinations, regions of great attraction for travel, such as business centres, industries, education centres, among others.

The increase in demand for transport, combined with short-term infrastructure plans that favour the car in recent years, have led to the unsustainable situation of congestion, accidents and environmental pollution that is common in large cities.

To resolve this issue, appropriate transport planning is needed, which aims to adapt a region's transport needs to its development according to its structural characteristics, implementing new systems or improving existing ones (CAMPOS, 2007).

In an attempt to put these fundamentals into practice, the metropolis of Salvador will be analysed, proposing a fictitious metro line to meet the city's demand for passenger transport in the years 1995, 2015 and 2025.

The city is the third most populous municipality in Brazil and the country's first capital. Due to its growth, in the 1990s it was decided to build two metro lines to make up for the city's transport shortages. The bidding process for the construction of the lines began in 1997, but due to numerous delays, the first section of Line 1 was only completed in 2014.

The base year for this project will be 1995, just like that of the metro system already planned and executed in Salvador. In this way, it will be possible to compare the product resulting from this work with that used in the metro lines built and under design.

A metro network with good operational planning, accessibility and comfort can be a factor that drives the migration of demand from car users to the metro. To do this, we will analyse whether the current metro network in the municipality of Salvador is really one that would serve car users.

1.2   Objectives

This research has two main objectives:

To analyse the demand for private and public transport for the implementation of a metro network in Salvador for the base year of 1995 and projected for 2015 and 2025.To propose a metro network for the city based on the lines of desire for private and public transport.

1.3   Work structure and method overview

The main methodological steps are briefly described below. The chapter of the text where each stage is detailed is also described.

1.3.1   Characterisation of the Study Region

Once the project's objectives have been clarified, the study region should be presented by contextualising the area in order to better understand the historical context of the city's urbanisation. Firstly, the historical development of Salvador and the Metropolitan Region will be discussed, explaining which processes have guided its occupation, and then a brief overview of the region's jobs will be given. The characterisation of the region will be described in section 3.1.1.

1.3.2   Data processing

The aim of this phase for transport planning is to define the pattern of travel and land use in the study area and to make a diagnosis of the existing transport system (CAMPOS, 2007).

The zoning adopted for the region from the Origin Destination Survey carried out in 1995 will be presented. Socio-economic data for the region was also obtained for the same year.

This data will be used to analyse the socio-economic characteristics of the region, an essential factor for predicting the movements of Salvador's inhabitants. This study will provide data for the next stage of calibrating the transport demand models. The geographical unit considered will be the traffic macro-zones (MZ).

Finally, the centroids of the macro-areas will be extracted and the most heavily loaded desire lines will be presented for the base year of 1995. In this way, it will be possible to analyse the zones with the greatest attraction and production of trips according to mode of transport, public and private motorised. This stage will be discussed in more detail in section 3.2.1.

1.3.3   Demand Model

The Four Stage model has the following phases: Trip Generation, Trip Distribution, Modal Splitting and Flow Allocation. In this work, models of the first and second stages of the sequential model will be calibrated.

In this phase, travel generation models are determined from the base year (1995). Finally, the values of trips produced and attracted by macro-areas are projected for 2015 and 2025. Based on these, the distribution of journeys for the three years considered will be carried out.

The result will be lines of desire for future years and maps showing in which regions macro-areas of greater attraction and travel production are concentrated. This methodological stage will be described in sections 3.2.2 and 3.2.4 and the results presented in sections 4.1 to 4.4.

1.3.4   Evaluating Solutions

Finally, a proposal will be made for a metro line that satisfies car demand for 2015 and 2025. The project will be compared to existing metro lines in the Salvador Metropolitan Region (RMS). The evaluation of the solutions will be covered in section 3.2.74.6.

CHAPTER 2

Theoretical Reference

2.1 Transport demand analysis

The demand for transport is the desire of some institution (person or group of individuals or companies) to move something from one place to another. The demand for transport is a consequence of other needs, such as work, study and leisure activities. So the demand for transport derives from the demand for other activities (KAWAMOTO, 2010).

According to Kawamoto (2010): "Transport demand analysis is the process by which one seeks to understand determinants of demand and how they interact and affect the evolution of traffic volumes."

The results of demand analysis are called transport demand models and are generally mathematical models that relate transport demand measures, socio-economic activities and user characteristics (KAWAMOTO, 2010). In general, transport demand models can be direct or sequential.

2.2 Direct Models

Direct models seek to represent the demand for a given mode and reason for travelling between an origin and destination pair. As the number of zones increases, the origin-destination pairs become more and more numerous, making the method very labour-intensive. Therefore, its ideal application is for situations with a small number of traffic zones (KAWAMOTO, 2010).

Examples of direct models are: linear projection, geometric or exponential projection, projection using the logistic curve and demand elasticity analysis (CAMPOS, 2007).

2.3 Sequential Models

Sequential Models aim to simplify the problem by breaking it down into smaller parts.

It is assumed that first the person decides to carry out an activity, then they decide where they are going to carry out that activity. Then they choose the mode of travel and the route. The order of the choices is just an assumption intended to simplify the problem (KAWAMOTO, 2010).

The model used in the project will be the sequential or Four Stage Model, which comprises the phases of Trip Generation, Trip Distribution, Modal Split and Traffic Allocation.

For the project, the first two stages will be carried out, since the final objective is to propose a metro network based on the O/D matrices of journeys by public and individual transport.

2.3.1 Travel generation

Trip Generation is the determination of the number of trips attracted and produced by a traffic zone. The basic procedure of the generation model according to Campos (2007) corresponds to the sequence:

- Identification of the determining factors of the base year;
- Determining the model to be used;
- Model calibration;
- Projection of socio-economic data for the project year;
- Application of the calibrated model;
- Determining future trips produced and attracted.

In this work, Travel Generation (production and attraction of trips) will be modelled using the multiple linear regression technique, which considers the influence of one or more independent variables, acting simultaneously and reflecting on the total number of trips. This technique measures the influence of each factor acting separately and its association with other factors. The purpose of this analysis is to produce, based on the traffic, land use and socio-economic data described above, an equation of the following form (BRUTON, 1979):

$$Y = k + b_1X_1 + b_2X_2 + \cdots + b_nX_n \tag{1}$$

Where:

Y: is the dependent variable (the zonal measure of traffic in terms of movements of people or movements by mode and purpose);

$X_1 \ldots X_n$: these are the related independent variables, such as land use and the socio-economic characteristics of the traffic zones;

$b_1 \ldots b_n$: these are the estimated coefficients of the respective independent variables;

k is a constant whose purpose is to represent the portion of y that was not explained by the independent variables.

A typical regression analysis relates the current values of the independent variables $X_1 \ldots X_n$)

with those of the dependent variable, for all the zones in the region under study. The "Least Squares" statistical adjustment process is then applied, estimating the values of the regression coefficients (b_1 ... b_n) and the constant (k) that best fit the existing data (BRUTON, 1979).

The resulting regression equation is then solved using the estimated future values of the independent variables to obtain the value of the dependent variable $(Y$ - number of trips generated per zone) for the projection years used in the study (BRUTON, 1979).

2.3.2   Travel distribution

Trip Distribution consists of assigning a number of trips between each pair of traffic zones, creating an O/D matrix of future trips from the base year data and the projections of trips produced and attracted (CAMPOS, 2007).

In this study, one of the two traditional models, Fratar or Gravitational, will be used for the journey distribution stage. The tool used for the calculation will be the software resulting from student Felipe Calsavara's Scientific Initiation project, which is still under development at the Transport Engineering Department of the São Carlos School of Engineering (University of São Paulo).

In the Fratar model, the forecast of future travel volume between a pair of zones is made by multiplying the current volume by the product of the growth factors forecast for the two zones, with adjustment for the relative attractiveness of the other zones (KAWAMOTO, 2010). The steps in building the model are as follows:

1   Having the trips attracted and produced by each zone for the base year and for the projected years, the adjustment value Li is determined for each zone of origin i;
2.   The volume of journeys distributed from zone i must be calculated using the equation:

$$Q_{ij}^1 = Q_{ij}^0 * F_i * F_j * L_i \qquad\qquad (2)$$

Where:

Q_{ij}^1 number of journeys in year t from i to j;

Q_{ij}^0 number of current journeys from i to j;

F_i growth factor of the macrozone of origin i;

F_j growth factor of the destination macro-zone j;

L_i source adjustment factor.

3   Once the flows have been calculated for all the origin-destination pairs, the rows and columns should be added together and the results compared with the predicted values for year t. If there are differences greater than the values initially stipulated for some of the rows or columns,

the process should go through a second iteration, taking the values determined in item 2 as the origin-destination matrix (KAWAMOTO, 2010).

The Gravitational Model adapts Newton's concept of gravity and is based on the assumption that journeys between zones are directly proportional to the attraction of each zone and inversely proportional to a spatial separation function between zones (BRUTON, 1979). The model's equation has the following form:

$$t_{ij} = k \frac{P_i A_j}{R_{ij}^c}$$

(3)

Where:

t_{ij} : number of *journeys from i to j;*

k,c: parameters to be calibrated using data from the base year, c varies between 0.6 and.5;

P_i : total number of journeys produced by zone i;

A_j : total number of trips attracted by zone *j;*

R_{ij} : impedance variable between zones i and *j.*

Most models use a friction function (f_{ij}), defined as:

$$f_{ij} = \frac{1}{R_{ij}^c}$$

(4)

In these cases, the initial expression t becomes:$_{ij}$

$$t_{ij} = k P_i A_j f_{ij}$$

(5)

An impedance means any kind of opposition to movement and can be defined by one variable or by a set of variables such as distance, journey time and transport cost. When impedance is defined by a set of these variables, it is called a generalised cost (CAMPOS, 2007).

The advantage of this model over others is that it takes into account, in addition to attraction and production, the effect of spatial separation or ease of iteration between regions defined by the impedance function (CAMPOS, 2007).

2.   3. 3 Modal split

The Modal Split aims to determine the number of journeys by each mode of transport between traffic zones.

Following on from the travel distribution process, modal split models are used to "split" the O/D travel matrix into O/D matrices by mode of transport (CAMPOS, 2007). In the project in question, the O/D matrices were already defined by mode, public and private, and walking was excluded from the analysis. This phase was therefore not necessary.

Modal split models are divided into deterministic, such as linear regression, cross-classification and deviation curves; and probabilistic models, such as discrete choice models.

2.3.3 Traffic Allocation

Traffic Allocation assesses the distribution of travel flows on existing transport systems and new alternatives. The models used seek to assess the road system's ability to absorb the flow of trips generated by road transport (CAMPOS, 2007).

This stage will not be carried out in this project, as no road network was obtained for the region for the base year, and mainly because the aim of the project is only to present an alternative metro network that meets the needs of the region.

The models used in Traffic Allocation can be: "all or nothing", dispersion curves, allocation techniques with capacity restrictions, incremental traffic allocation techniques, multi-path allocation techniques, static equilibrium models and dynamic allocation models.

2.4 Lines   of desire

To find out how the demand for transport is distributed within the study area, you can use desire lines, which represent the volumes of journeys between an origin and destination pair. In this way, a map of desire lines makes it possible to compare demand between zones in the study region and determine which origin-destination pairs have the highest priority.

CHAPTER 3

Materials and Methods

3.1 Materials

3.1.1 Characterisation of the Study Region

The city of Salvador covers an area of 692.819 km^2 , according to the IBGE the estimated population for 2015 was 2,921,087 inhabitants.

Salvador was built on the shores of the Bay of All Saints and grew from the harbour. The presence of hillsides and the hilly terrain directly interfered with the urbanisation process, initially dividing the city into two parts: the upper part, where Praça Municipal and Praça Castro Alves are located today, and the lower part, where Mercado Modelo, the Port and Comércio are located (MENDES, 2006).

At the beginning of the 20th century, the city's historic centre had modes of transport adapted to bridge the gap between the lower and upper town. There were tram systems, lifts and inclined planes that satisfactorily met the inhabitants' transport needs.

From the 1950s onwards, there was an excessive increase in the density of occupation in Salvador's traditional neighbourhoods. These neighbourhoods had basic infrastructure and services, so the price of land was high, which led to the expulsion of the lower classes to more remote areas that lacked infrastructure.

Today, the traditional centre is gradually being degraded as it proves incompatible with the rapid growth of the car fleet, as well as becoming unattractive to real estate interests (CARVALHO et al., 2014). In the 1970s, a second centre was built, Centro do Iguatemi, symbolising modernity with a new bus terminal and shopping centre. This area became consolidated as a place with a high concentration of traffic flow, which later led to saturation of the area's internal and external roads.

At the same time, the city began to extend towards the Atlantic coast with the expansion of the road system through the opening of Avenida Paralela (Av. Luis Viana), setting up a vector for south-north expansion and connecting spaces that had been empty until then, but already appropriated by property developers (CARVALHO et al., 2014).

Expansion into the northern region occurred in two real estate patterns: with the establishment of housing estates for the so-called "lower middle classes" in the interior of the municipality, in the Miolo Sul and Miolo Norte regions, and high-end residential subdivisions and horizontal and vertical condominiums on the Atlantic coast, in the Orla Sul and Orla Norte regions. The appreciation of the waterfront and the cost of urban land in the city have pushed the low-income population to the geographical centre of the municipality, to the edges of Todos os Santos Bay, in the Railway Suburb, and to some municipalities in the metropolitan region (CARVALHO et al., 2014).

These interventions, property investments and the development of the road structure have consolidated three vectors of expansion. The North Waterfront represents the high-end region, where wealth, tourism, job opportunities and income are concentrated. The Miolo, which is located in the geographical centre, is made up of the lower middle class and has very hilly terrain with the presence of several popular allotments and invasions.

The Railway Suburb, facing the Bay of All Saints, is characterised by several popular housing developments that were encouraged from the 1940s onwards.

Figure 1 shows the main transport corridors in the city of Salvador, BR 324 (in green) and Av. Luis Viana (in orange), axes on which part of Line 1 was built and Line 2 of the current Metro is planned. Also shown are the regions of the city dealt with earlier: the Centre, Orla Norte, Miolo and Subúrbio Ferroviário.

The Iguatemi Centre area is located at the junction of BR 324 and Av. Luis Viana and is a very problematic traffic flow area, with intense congestion, especially at peak times.

Figure 1: Salvador's main roads and development axes Source: (Google Earth, 2008)

Figure 2 shows a schematic of the development of the city of Salvador from 1940 to 1980, showing how the area was occupied along the growth vectors presented and which are consolidated today, as can be seen in the satellite image presented earlier in Figure 1.

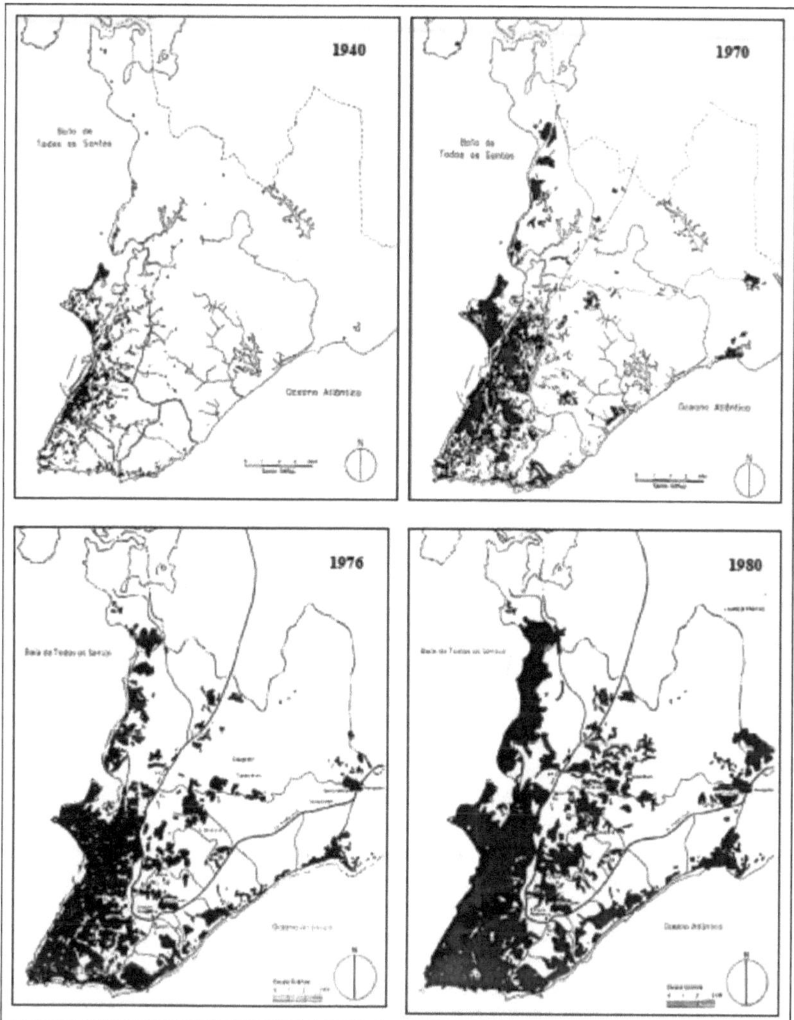

Figure 2: Evolution of Salvador's Urban Area, from 1940 to 1980, Source: (GORDILHO-SOUZA, 2000)

3.1.2 Base Data

The basic data used in the design of this project comprise:

-Macrozoning used in the Origin/Destination survey carried out in 1995;

-Summary of socio-economic and physical data by macro-zone from the Origin/Destination survey carried out in 1995;

-Origin/Destination matrices of a sample, corresponding to public, private and walking modes of transport, comprising total periods (for all modes) and peak periods (only for public transport), for the base year 1995;

-Population and average household income data   obtained from the Brazilian Institute of Geography and Statistics (IBGE);

-Disaggregated Origin/Destination Survey 2012, from which the total number of jobs in the municipality was extracted;

-Information on the   route of the current metro line, obtained from the *website of* CCR (Companhia de Concessões Rodoviárias), the concessionaire for the stretch.

3.1.3   Applications Used

The following applications were used to design this project:

-*TransCAD   - Transportation Planning Software - Caliper Corporation: an* application that combines Geographic Information System (GIS) and transport modelling resources. It was used to generate thematic maps;

-*Google   Earth:* satellite image visualisation application. It was used to generate maps, analyse the proposed line and generate its altimetric profile;

-SPSS *22.0 - IBM:* statistical application. It was used to calibrate the trip generation models.

-*Grav   Tar 1.0:* programme resulting from the Scientific Initiation project of student Felipe Calsavara, still under development in the Engineering Department
of Transport at the São Carlos School of Engineering (University of São Paulo). The application was created using *MATLAB - MathWorks.*

3.2   Method

3.2.1   Data processing

3.2.1.   1Zoning

In order to simplify the study, traffic zones are created, geographical units that allow data to be grouped in such a way as to make them manageable (BRUTON, 1979). Traffic zones are areas with similar characteristics in terms of income, density, occupation and travel patterns, delimited by geographical features or regions with different characteristics.

This project will use the 63 macro-areas delimited in the 1995 Origin-Destination survey. Macrozones are groups of traffic zones, aggregating regions with similar characteristics. Sub-regions are aggregations of neighbourhoods and large regions with similar characteristics. Figure 3 illustrates Salvador, its macro-zones and sub-regions.

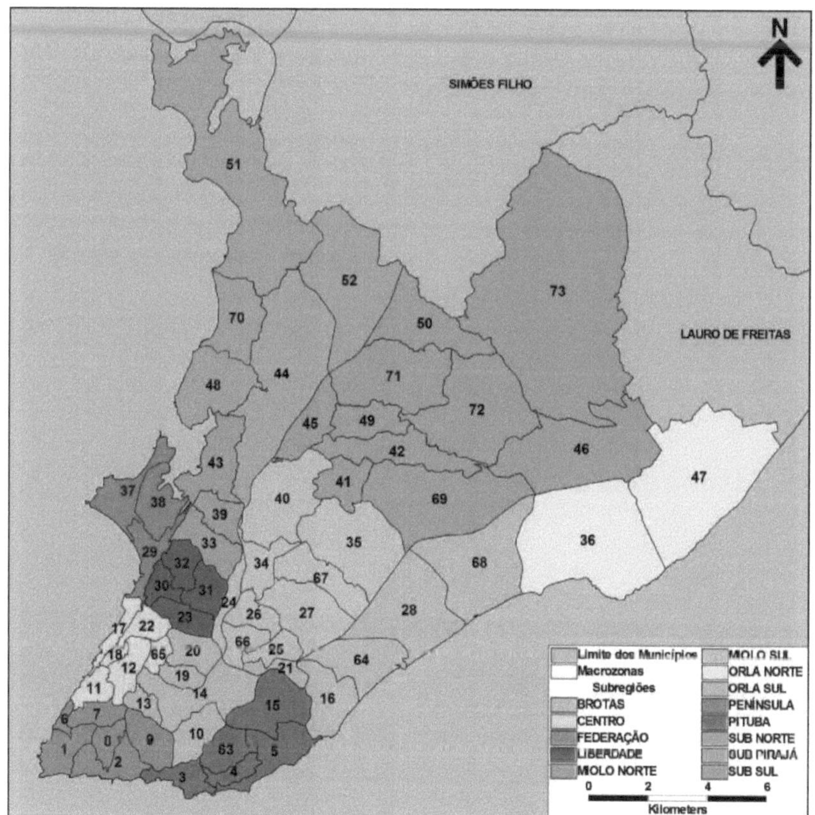

Figure 3: Representation of Salvador divided into macro-zones and sub-regions

By delimiting the macro-zones, it is possible to extract the centroids, which symbolically consist of the points of origin or destination of trips and should be seen as the centre of gravity of the macro-zones. Therefore, in the mathematical simulation models, the zones do not exist as an area, but are represented by the centroids; the area representation is merely illustrative to facilitate understanding and analysis of the data.

This project relies on data from the OD Survey carried out in 1995. The Household Survey contains data on the socio-economic characteristics and travel behaviour of the macro-areas. The data obtained was: population;

average income per household; households with none, one, two or more vehicles; total number of vehicles; total number of households; employed population; students; people per household; people by age group (0 to 4 years, 5 to 13 years, 14 to 21 years, 22 to 44 years, 45 to 60 years, over 60 years); total jobs generated (by sector); total trips produced (morning peak, between peak and afternoon

17

peak) and total trips attracted (morning peak, between peak and afternoon peak).

2.3.3.1 Analysing Socioeconomic Data

By analysing the socio-economic data provided, it is possible to consolidate the information presented in the previous section on the city's urban development.

The total population of the macro-areas is a significant factor in estimating trip generation. Figure 4 shows the distribution of the population (O/D Survey sample) in 1995.

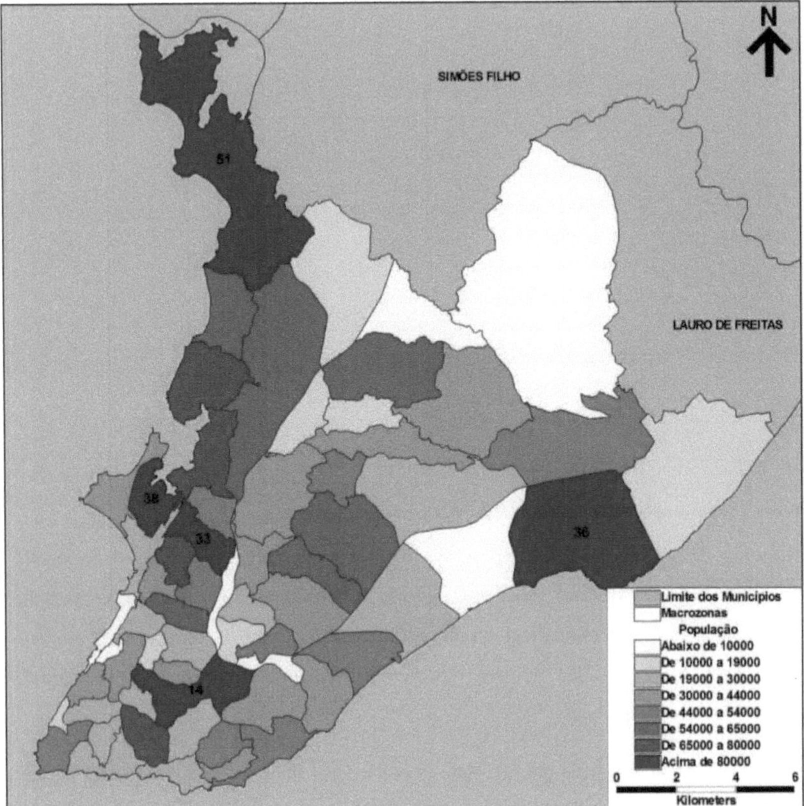

Figure 4: Population by Macrozone

The regions of Brotas (MZ 14), Coutos (MZ 51), Malvinas (MZ 36), Liberdade (MZ 32), Fazenda Grande Retiro (MZ 33) and Massaranduba (MZ 38) stand out. The map corroborates the analysis of the city's growth vectors: the large number of inhabitants near the railway line, in the Orla Norte region and near the meeting of the BR 324 and Av. Luis Viana corridors in Brotas.

Figure 5 shows the average income per household in R$ (reals) by macro-area, which is important information as it affects the number, frequency and mode of transport used by residents.

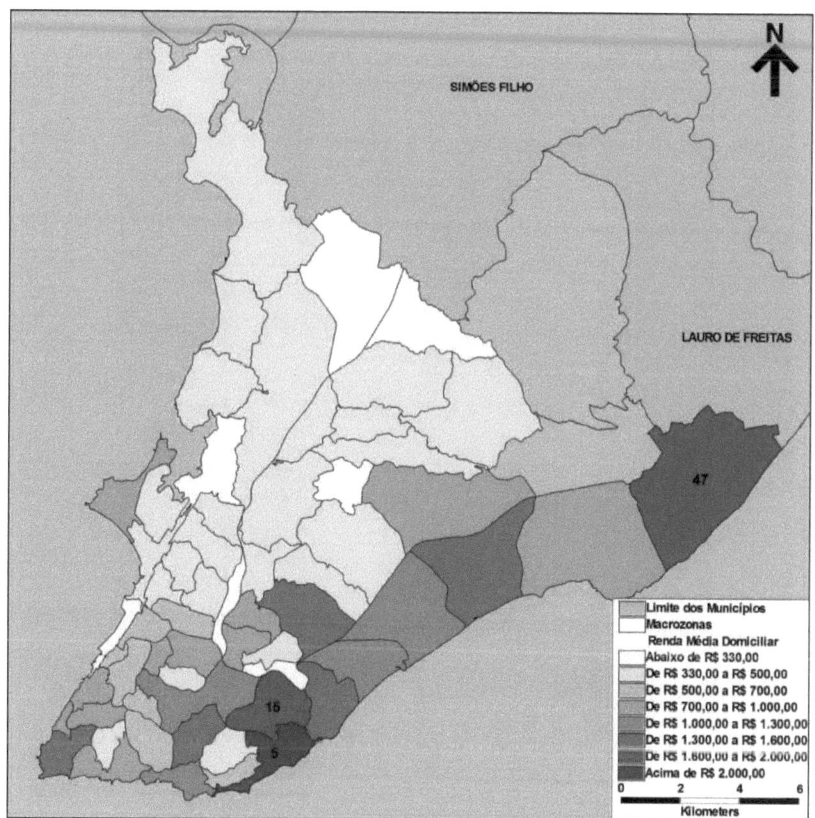

Figure 5: Average Income per Household (R$) by Macrozone

It can be seen that the Stela Maris region (MZ 47) on the northern fringe, Iguatemi (MZ 15) and Pituba (MZ 5) stand out as the macro-areas with the highest purchasing power. The average income per household was R$ 688.16 for the municipality, with a high standard deviation of R$ 457.75.

Figure 6 shows the vehicle per household ratio for each macro-zone, indicating in which zones the tendency to travel by private transport is greater. It can be seen that the possession of a greater number of cars per household is closely linked to a higher income.

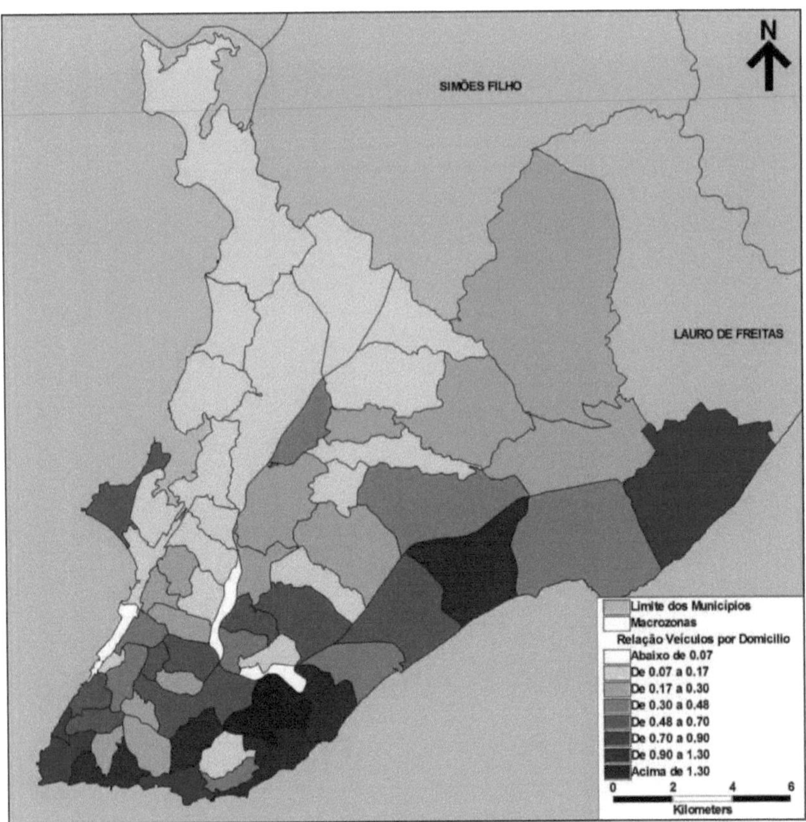

Figure 6: Vehicles per household by Macrozone

It can be seen that the same macro-areas that stand out on the income map also stand out on the map of vehicles per household. The average number of vehicles per household was 0.36 for the municipality, with a high standard deviation of 0.31.

Figure 7 shows the number of jobs generated in each macro-zone, an essential item when analysing travel attraction.

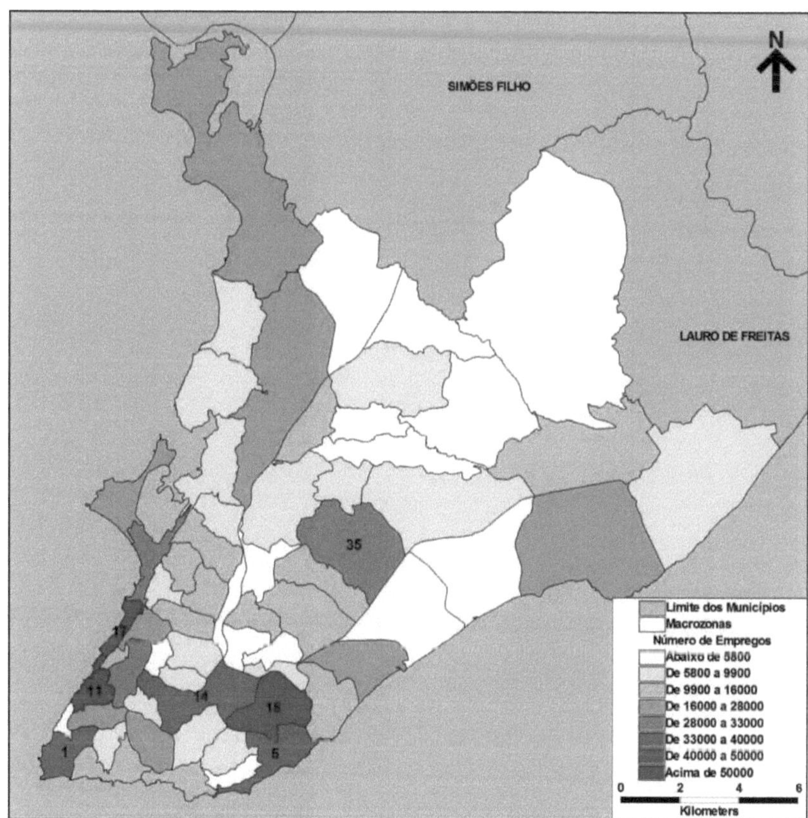

Figure 7: Number of Jobs Generated by Macrozone

The regions of Iguatemi (MZ 15), Campo Grande (MZ 11) and Comércio (MZ 17) stand out, while Brotas (MZ 14), Pituba (MZ 5), Barra (MZ 1) and Arenoso (MZ 35) are less important. It can be seen that jobs are concentrated in central areas and those with greater purchasing power. Arenoso, which belongs to the Miolo Sul sub-region, is home to the CAB (Bahia Administrative Centre), which generates many jobs.

The average number of jobs in the municipality was 13382, with a high standard deviation of 11672, so the distribution of jobs between the city's macro-areas is very uneven, as is the case with the previous items.

Presenting a map with the predominant type of employment by macro-zone is not pertinent, since almost all the zones had a predominance of jobs in Commerce.

Figure 8 shows the percentage of the population occupied by macro-area.

21

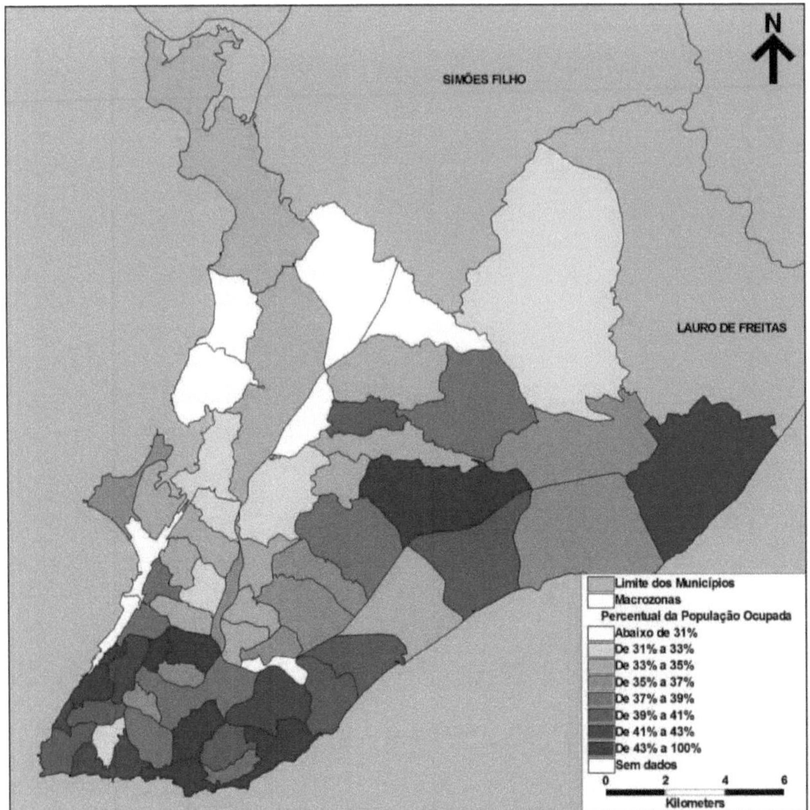

Figure 8: Percentage of Occupied Population by Macrozone

It can be seen that areas with lower average household incomes have a lower percentage of occupied population, as can be seen in the Subúrbio Ferroviário and Miolo. The average occupied population in the municipality is 36.06%, with a standard deviation of 6.65%.Table 1 summarises the socio-economic characteristics.

Table 1: Summary of the region's socio-economic characteristics

Features	Average	Standard Deviation
Population	37533,21	23761,47
Average Household Income	700,46	457,75
Vehicles per household	0,36	0,31
Number of Jobs Generated	13381,35	11671,92
Occupied Population in relation to Total Population	36,16%	6,65%

2.3.3.2 Origin/Destination matrices

The O/D matrices for public transport (total and peak) and private transport (total) were

22

provided. For demand forecasting, the peak and inter-peak period matrices are used in the case of an operational analysis, which is not the case here.

As a result, the total private motorised mode matrix had to be adapted for the peak period. To do this, an analogy was made with the public transport matrices, obtaining the percentage of the total matrix that corresponds to the peak period. The total private mode matrix was then multiplied by this percentage to obtain the peak hour matrix.

3.2.2 Travel Generation Modelling

In the Trip Generation phase, the trips attracted and produced by macro-zone were estimated. As presented in subsection 2.3.1, the multiple linear regression technique will be used for this stage.

IBM SPSS 22.0 statistical software was used to estimate the linear models of trips produced and attracted by public and private motorised transport modes. The 1995 database was used to estimate the parameters of the models.

Usually, the main independent variables considered in the production of trips are: income, vehicle ownership, number of households, number of people employed, population or population density, number of people of school age. As for attracting trips, the following are taken into account: area destined for industry and commerce, number of jobs, school enrolment and accessibility (CAMPOS, 2007).

For the models developed in this project, the variables population and average household income were used for trips produced and jobs for trips attracted.

Some parameters of the regression model were evaluated to assess its applicability, such as residuals, coefficient of determination R^2 , sum of total squares, residuals and regression, t-statistic and significance test of the estimated parameters.

A residual or deviation (ε_i) is defined as $\varepsilon_i = Yi - \hat{Y}_i$ where Yi is a real observation of the variable Y and $\hat{Y}_i$ is the value estimated using the regression model. The residuals of a regression give a measure of the quality of the fit, if the residuals are large the fit is bad, if they are small the fit is good (BARBETTA, 2012).

The total sum of squares is given by evaluating only the arithmetic mean, without taking into account the relationship between X and Y (BARBETTA, 2012):

$$SQT = \Sigma(y - \bar{y})^2 \qquad\qquad (6)$$

If the different values of X *are* taken into account by the regression equation, all that remains is what the regression equation cannot explain, i.e. the sum of squares due to random error (BARBETTA, 2012):

$$SQE = \Sigma(y - \hat{y})^2 \qquad\qquad (7)$$

The difference between *SQT* and *SQE* is known as the sum of squares of the regression *(SQR)*. *It* is the part of the variation in Y that the regression equation is able to explain more than simply the arithmetic mean of Y (BARBETTA, 2012).

The ratio between *SQR* and *SQT* is called the coefficient of determination:

$$R^2 = \frac{SQR}{SQT} = \frac{variação\ explicada}{variação\ total} \qquad\qquad (8)$$

The coefficient of determination, or R^2 , can be interpreted as a descriptive measure of the proportion of the variation in Y that can be explained by X, according to the model adopted. The closer it is to 1, the more the variability of Y is explained by variable X (BARBETTA, 2012).

The *t-statistic* corresponds to the Student's t-test, a hypothesis test that rejects or rejects a null hypothesis when the test statistic (t) follows a Student's t distribution. This distribution is relevant because it uses the sample estimate of the error variance, rather than its true value (BARBETTA, 2012).

The hypothesis must be formulated, the value of í shown below in equation (9) must be calculated, and it must be applied to the probability density function of Student's t distribution, measuring the size of the area below this function for values greater than or equal to t. *This area represents the average probability of this sample having presented the observed value or something extreme (BARBETTA, 2012).* This area represents the average probability of this sample having presented the observed value or something more extreme (BARBETTA, 2012).

$$t = \frac{\bar{x} - \mu o}{\left(\frac{s}{\sqrt{n}}\right)} \qquad\qquad (9)$$

If the probability of this result occurring is small, it can be concluded that the observed result is statistically relevant.

The *t-test* can also be used to test the significance of regression coefficients. This test is used to confirm that the variable being used is actually contributing to the estimate (BARBETTA, 2012).

If the significance is greater than 5% (less than 95% confidence) the variable can be removed from the model without any major loss in quality.

2.3.4 Projecting production and attracting future trips

In order to project the future values of travel production and attraction, it is necessary to project the independent variables chosen for the future years analysed.

The projections of these variables were based on data from the 1995 Origin/Destination survey by the Brazilian Institute of Geography and Statistics (IBGE) and the 2012 disaggregated Origin/Destination survey.

To estimate average household income, the minimum wage was adjusted to equalise the value of the currency for the periods analysed. The adjustment rate was found by analysing the growth of the minimum wage between 1995 and 2010.

In this way, the corrected average household income for 2010 was obtained by dividing the income figure provided by the IBGE by the adjustment rate found, as indicated by Rocha (2014).

These figures were used to project future journeys. The data was presented in the form of maps, which better illustrate the contrasts between the results for each macro-zone.

3.2.4   Travel distribution modelling

The Trip Distribution phase involved assigning a number of trips between each pair of traffic macro-areas. In this stage, two traditional models were used, Fratar and Gravitational, already presented in subsection 2.3.2, which will be analysed and compared later.

In the Fratar model, the forecast of future trips (Q/y) between a pair of zones is made by multiplying the current volume (Q?-) by the product of the growth factors *(F_t and f})* forecast for the two zones, with adjustment *(Li) for* the relative attractiveness of the other zones, as shown in expression (2).

In the Gravity model, the prediction of future trips *(Vij)* is directly proportional to the product of the trips generated in the origin zone $(P_t$) and those attracted in the destination zone (/>■), and inversely proportional to the distance (d$_{i;}$) between them, as shown in expression (10). There are also two factors *(k and a)* that are obtained from the calibration of the model, such as linear calibration.

$$V_{ij} = \frac{k*P_i*P_j}{d_{ij}^{\alpha}} \tag{10}$$

In the Gravitational model, the spatial separation of the macro-zones was considered as an impedance factor.

The distance matrix considered intra-zonal distances, i.e. the distances between journeys made within the same zone, which are different from zero. The method proposed by Plaza (2014) was used to determine these distances.

According to Plaza (2014), macrozones can be separated into two groups, with areas smaller and larger than 10 km^2 .

For small areas, the global distance of intrazonal displacement is equal to:

$$D_G = 0{,}5648 * A^{0{,}4558} \qquad\qquad (\,11\,)$$

For large areas, the global distance of intra-zonal displacement is equal to:

$$D_G = 0{,}6882 * \ln(A) - 0{,}1319 \qquad\qquad (\,12\,)$$

With the O/D matrices for the base year, the trips produced and attracted projected for future years and the distance matrix, the trip distribution can be calculated using the two models presented. The tool used for the calculation was the code resulting from student Felipe Calsavara's Scientific Initiation project, which is still under development at the Transport Engineering Department of the São Carlos School of Engineering (University of São Paulo).

The code resulted in an error if there were null values in the matrices. Therefore, in the journeys matrix, all values equal to zero were considered to be 0.001, the smallest possible amount that the application allowed.

3.2.5   Obtaining travel flows and desire lines

After calculating the distribution of trips using the two models, it was necessary to choose which would be the most appropriate to reflect the region's growth.

We chose to compare histograms that related the distance between origin and destination pairs and the number of trips, analysing their shape and the distribution of trips. The histograms of trips from the base year were compared with histograms estimated by the Fratar and Gravitational models also for the base year, i.e. the input data for the models was the O/D matrix for the base year and the trips produced and attracted estimated by the trip generation model for the base year. In this way, the calibrations with the most similar distribution to the base year were chosen.

Once the most appropriate O/D matrix for future years has been adopted, the information can be consolidated using desire line maps, which represent the main flows in the region.

As the desire lines represent the volume of journeys for each pair of origin and destination, the result of the map is a "tangle" of lines. We therefore chose to select only the shipments above the 95th percentile, i.e. the top 5% of the sample.

3.2.6   Current Line Assessment

The current line was evaluated by analysing the results of the previous phases.

The projection of trips produced and attracted by mode was used by macro-zone to assess whether the macro-zones with the greatest production and attraction were being served by the current line. The desire lines were used to assess whether the main flows in the region were being served. Not only the stretches currently built were considered, but also the stretches under construction and planned, as available on the CCR Metrô Bahia website.

3.2.7   Preliminary Metro Network Proposal

Like the evaluation of the current line, the results of the previous phases were used in the preliminary proposal. The maps produced were analysed in order to propose lines that would make up for the shortcomings of the current line and to propose lines that would also serve car users.

In proposing the line, it was considered that the metro would be underground and that the soil would be ideal, since there is no geotechnical data and the complexity of the project would increase greatly if it went into this. It is worth noting that a metro network is proposed based only on analysing the demand for passenger transport. It is recommended that future work assess the feasibility of the project, taking into account technical, construction, environmental and financial criteria.

Using the "Elevation Profile" tool in the *Google Earth software,* it was possible to assess the relief of the area where it was to be built. It was observed that, due to the city's hilly terrain, the lines should ideally follow the valley avenues already consolidated in the city.

Stations were proposed along the new network. Large sites, such as squares, were chosen for them. These sites were named after the neighbourhood in which they were located, giving them the names of neighbourhoods, squares and parks.

CHAPTER 4

Results and Discussions

4.1 Trip generation model

Table 2 shows the model of journeys produced by public transport, based on population. The population variable was used because it is closely related to the number of trips produced in the macro-areas and the model resulted in reasonable values. The constant was excluded from the model because it was significant at approximately 68 per cent, so the variable can be removed from the model without a major loss in quality.

Table 2: Model of journeys produced by public transport

Model		Independent Variables			
R^2	0,964		Coef	t	Sig.
Sum of Squares					
Regression	3749369853,513	Const ant-			
Waste	138879760,487	Population0	.174	40,912	0,000
Total	3888249614,000				

The statistics described above can be observed. The sum of squares of the residuals was lower than that of the regression and the coefficient of determination was close to one, showing that the model fitted the data well.

The constant *a was* not used in this model, and the independent variable *e* explains the entire model. The *t* coefficient has a high value and significance corresponds to 0%, which means that the estimated parameter is significantly different from zero.

Table 3 shows the model of trips attracted by public transport, based only on the number of jobs generated in the macro-area. The variable of jobs generated by macro-area was chosen because it represents a large part of the reasons for journeys made.

A model considering the constant was tested, but found to have a negative value. This hypothesis was therefore discarded and the null constant model was adopted.

Table 3: Model of journeys attracted by public transport

Model		Independent Variables			
R^2	0,873		Coef	t	Sig.
Sum of Squares					
Regression	5734531855,523	Const ant-			
Waste	830431752,477	Employment0	.539	20,692	0,000
Total	6564963608,000	--			

The sum of squares of the residuals was lower than that of the regression and the coefficient of determination was high at 0.873, demonstrating a good fit of the model to the data. The *t coefficient has a* high value and significance corresponds to 0%. The constant was not added to the model as it resulted in a negative value.

Table 4 shows the model of trips produced by private motorised transport, based only on average household income. This variable was chosen because a relationship was observed between higher average household income and a higher vehicle per household ratio, shown in Figure 5 and Figure 6. When the model was tested, it was found that the trips produced by the private motorised mode were better explained by the average household income variable than by the population variable. The constant was discarded because it was negative.

Table 4: Model of trips produced by the private motorised mode

Model		Independent Variables			
R^2	0,682		Coef	t	Sig.
Sum of Squares					
Regression	280569506,218	Constant			
Waste	130667076,782	Median household income	2,528	11,538	0,000
Total	411236583,000				

The sum of squares of the residuals was lower than that of the regression and the coefficient of determination was 0.682, showing that the model fitted the data reasonably well. The *t coefficient is* high and the significance level is 0%.

Table 5 shows the model of trips attracted by the private motorised mode, based only on the number of jobs generated. As already observed in the model of trips attracted by public transport, the number of jobs generated in a macro-zone is one of the main reasons for attracting trips to it.The constant obtained had a negative value and was discarded in the final model.

Table 5: Model of trips attracted by the private motorised mode

Model		Independent Variables			
R^2	0,778		Coef	t	Sig.
Sum of Squares					
Regression	322418222,955	Constant			
Waste	92095602,138	Employment	0,128	14,733	0,000
Total	414513825,092				

The sum of squares of the residuals was lower than that of the regression and the coefficient of determination was 0.778, showing that the model fitted the data reasonably well. The *t coefficient is* high and the significance level is 0%. The constant was not added to the model as it was negative.

The Bus Station macro-area (MZ 21) had zero population, fleet and average household income. Due to these null values, the model resulted in different values than expected. It was decided to exclude it from the models, as the number of trips produced and attracted was negligible when compared to the others. In addition, the so-called null zones are a problem in the case of aggregate models of attraction and production of trips (ORTÚZAR, WILLUMSEN, 2011).

The final models used are presented below. In equation (13) POPULATION corresponds to population, in equations (14) and (16) EMPLOYMENT corresponds to the number of jobs and in equation (15) INCOME refers to average household income.

$$VProduced_{Collective} = 0.174 * P1PULATION \qquad (13)$$

$$Collective_{turnover} = 0.539 * EMPLOYMENTS \qquad (14)$$

$$VProduced_{Private} = 12.311 * INCOME \qquad (15)$$

$$Private_{VAtracted} = 0.623 * EMPLOYMENTS \qquad (16)$$

The growth rate of the minimum wage from 1995 to 2010 was analysed, corresponding to 173.33%, as shown in Table 6 below.

Table 6: Rate of increase in the minimum wage from 1995 to 2010 (Source: Regional Labour Court -TRT)

Year	Minimum wage	Rate of increase (%)
1995	100	0,00%
1996	112	12,00%
1997	120	7,14%
1998	130	8,33%
1999	136	4,62%
2000	151	11,03%
2001	180	19,21%
2002	200	11,11%
2003	240	20,00%
2004	260	8,33%
2005	300	15,38%
2006	350	16,67%
2007	380	8,57%
2008	415	9,21%
2009	465	12,05%
2010	510	9,68%
Total		173,33%

The corrected average household income for 2010 was obtained by dividing the income figure provided by the IBGE by the adjustment rate found (1.73). Using the values for 1995 and 2010, a straight line was obtained that passes through these points. Using simple linear regression, the following equations were obtained for the variables that feed the models:

$$POPULATION = -84863635,640 + 43637,476 * AN7 \qquad (17)$$

$$EMPLOYMENT = -41137877,352 + 21043,058 * YEAR \qquad (18)$$

$$INCOME = -10485.136 + 5.590 * AIO \qquad (19)$$

Using these models, it was possible to estimate the total number of trips produced and attracted by public and private transport modes for the years 2015 and 2025.

4.2   Projected future trips

The results of the trip generation stage are the total number of trips produced and attracted by mode for the years 2015 and 2025. In order to present the data more dynamically, thematic maps of the total number of trips by macro-areas were produced.

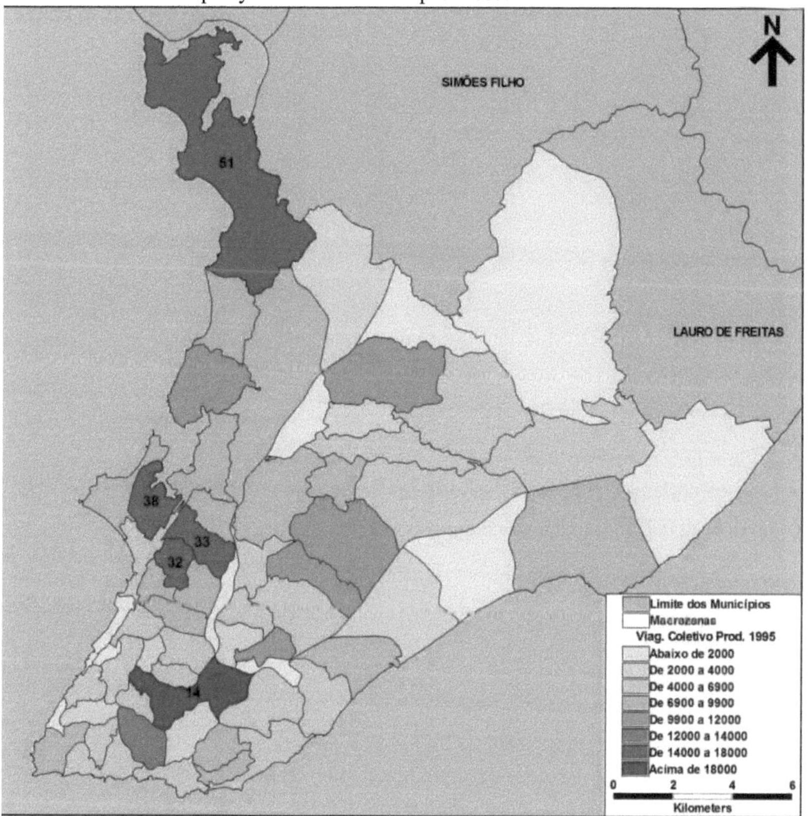

Figure 9: Collective mode journeys produced for 1995

It can be seen that, in the base year, the journeys produced are concentrated in the macro-areas of Brotas (MZ 14), Liberdade (MZ 32), Fazenda Grande Retiro (MZ 33), Massaranduba (MZ 38) and Coutos (MZ 51), regions with large populations, as seen previously in Figure 4.

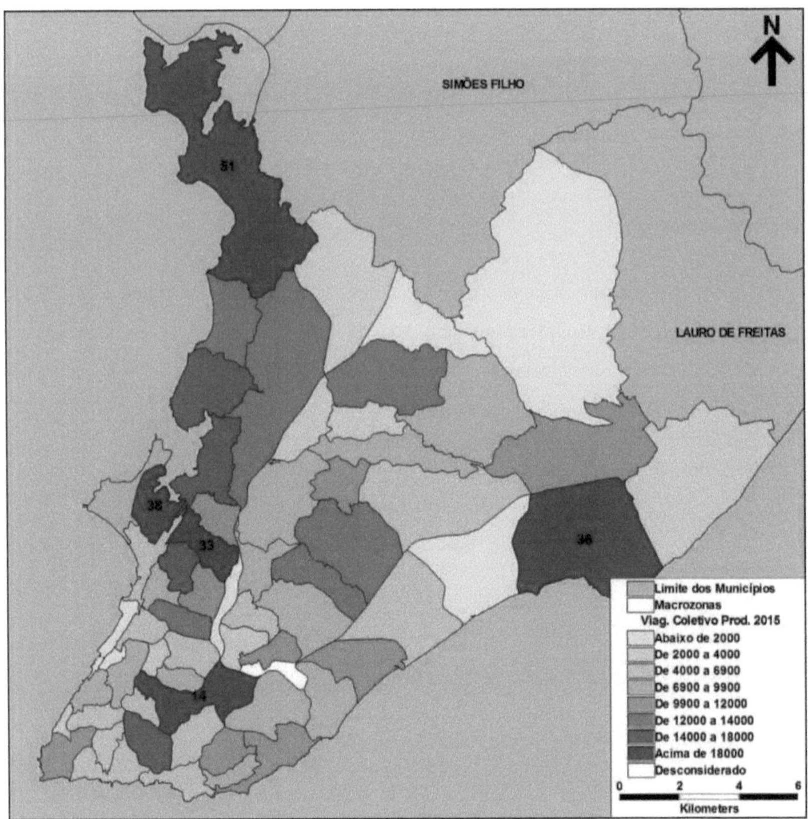

Figure 10: Collective Mode Trips Produced for 2015

In 2015, the macro-areas with the highest trip production are roughly the same as in 1995, with the inclusion of Malvinas (36) on the northern fringe.

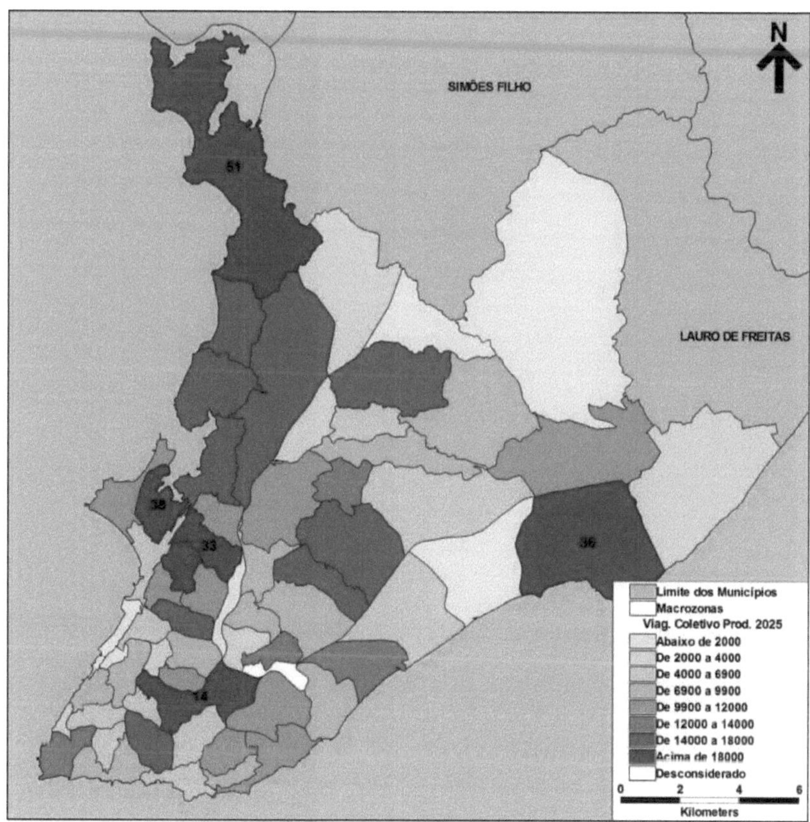

Figure 11: Collective mode trips produced for the year 2025

The total number of journeys produced by public transport in 2025 follows the trend already shown in the 2015 map.

The evolution of journeys attracted by public transport is shown in the following figures.

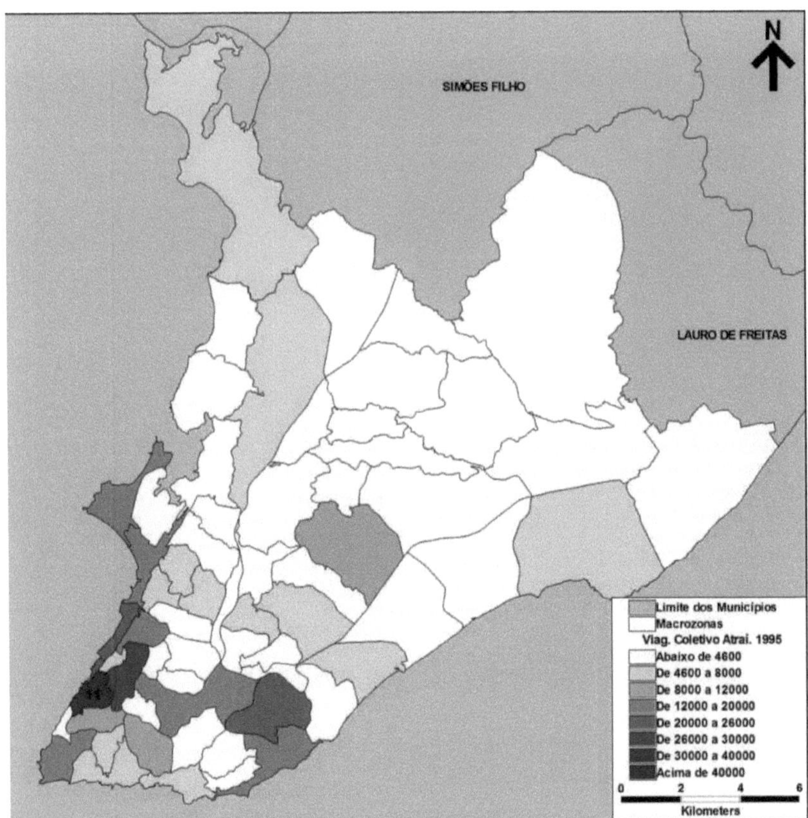

Figure 12: Collective Mode Trips Attracted for 1995

In the base year, only the Campo Grande macro-area (MZ 11) stands out visibly in relation to the others.

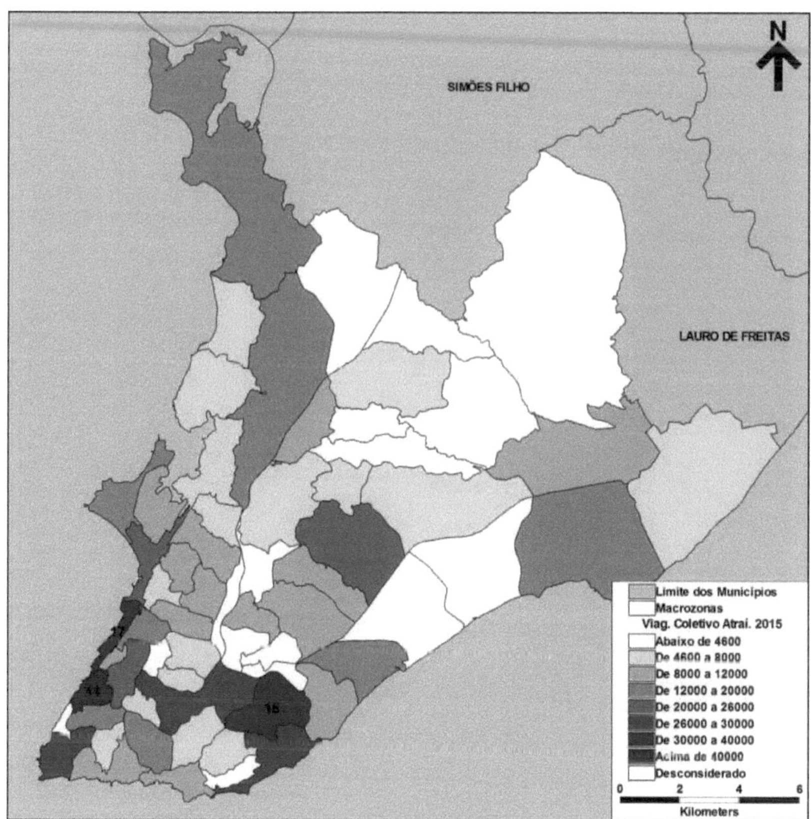

Figure 13: Collective Mode Trips Attracted for 2015

For the 2015 projection, it can be seen that in addition to Campo Grande (MZ 11), Iguatemi (MZ 15) and Comércio (MZ 17), near the Lacerda Elevator, stand out as regions with a large number of jobs, as can be seen in Figure 7.

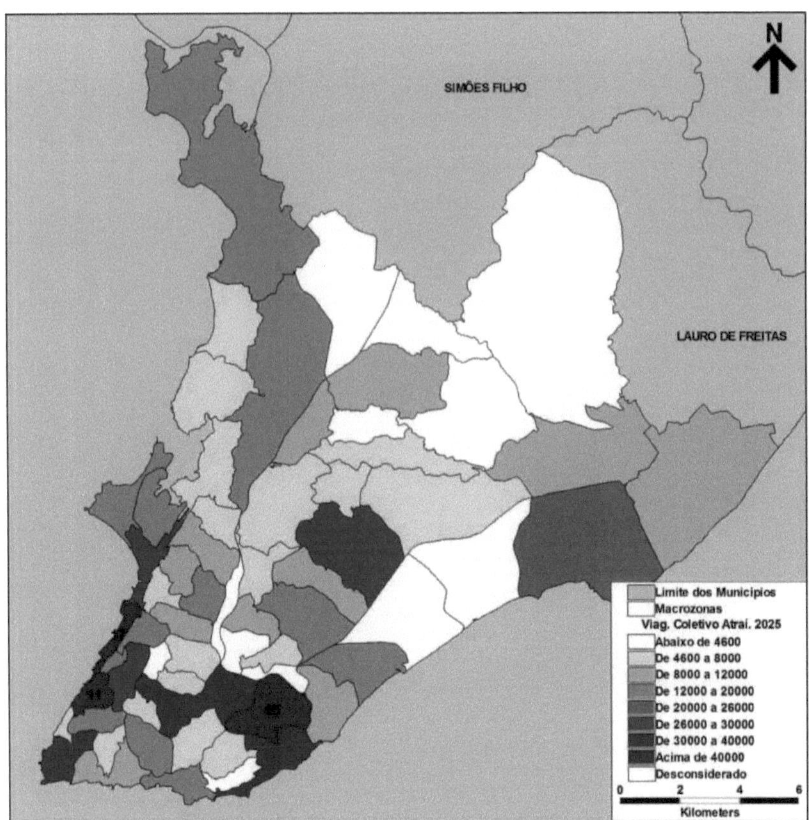

Figure 14: Collective Mode Trips Attracted for the Year 2025

The projection for 2025 follows the same trend as the 2015 map.

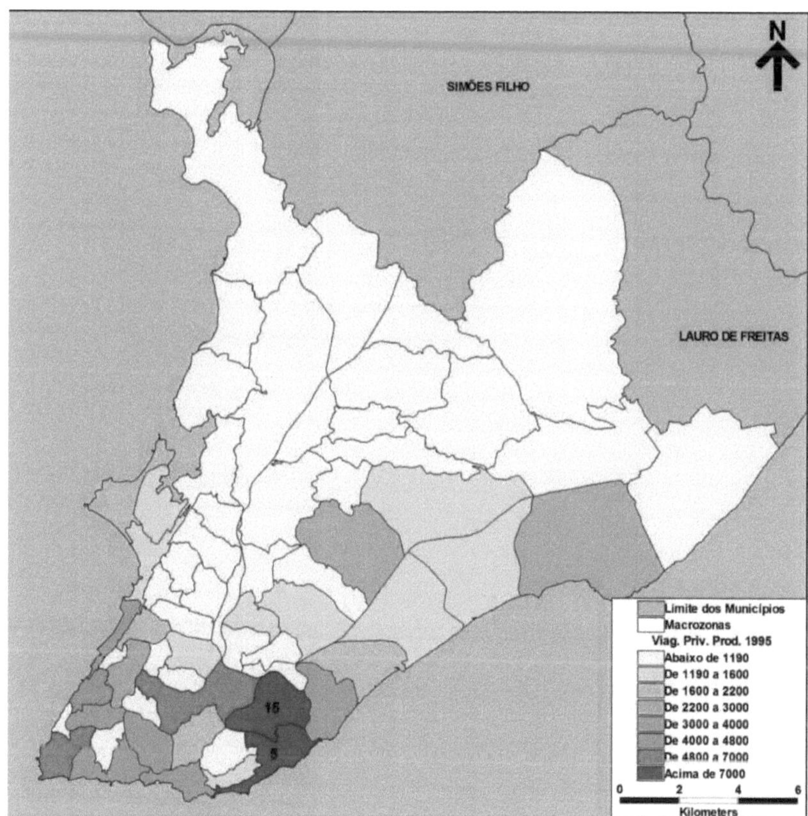

Figure 15: Private Mode Trips Produced for 1995

As for trips produced by private mode in the 1995 base year, the Iguatemi (MZ 15) and Pituba (MZ 5) macro-areas stand out, regions with a high vehicle per household ratio, as can be seen in Figure 6.

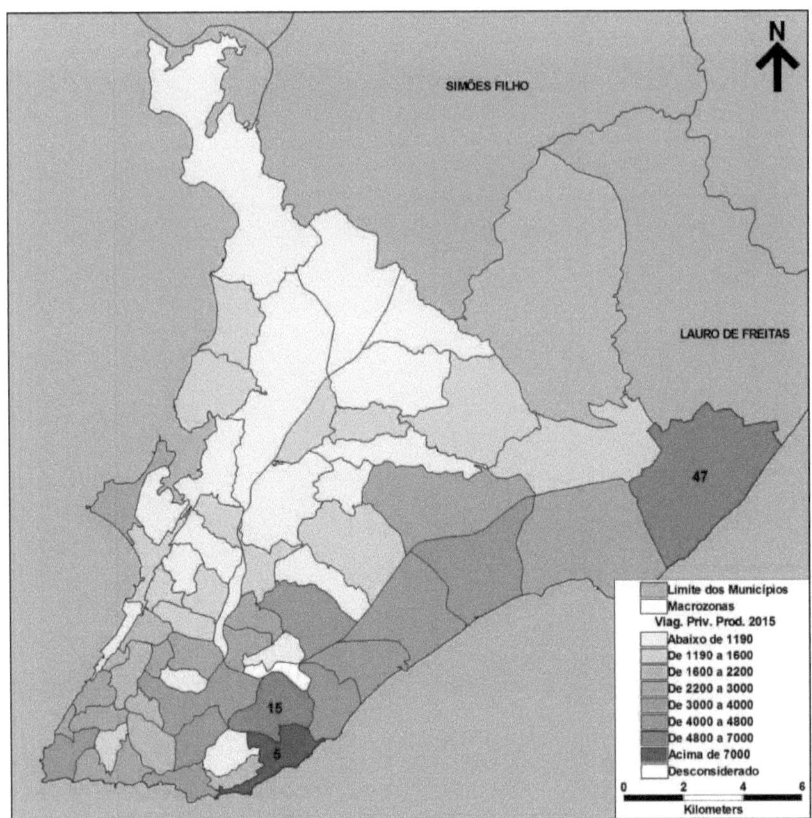

Figure 16: Private Mode Trips Produced for 2015

The trend in trips produced in 2015 is similar to that shown for 1995, with a new highlight in the Stela Maris macro-area (MZ 47) on the northern fringe, another region with a high vehicle-per-household ratio, shown in Figure 6.

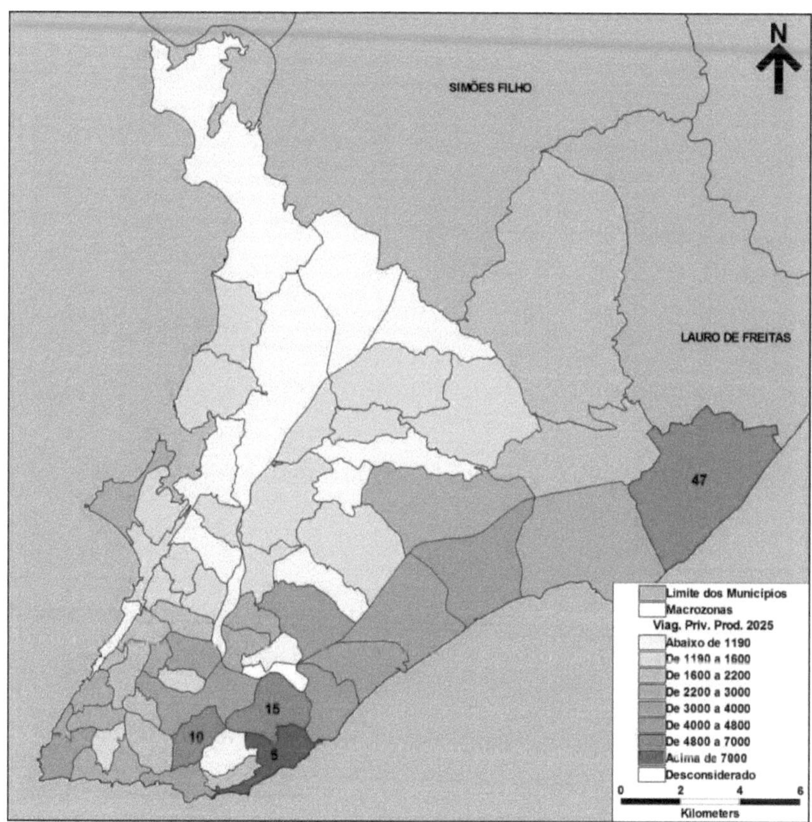

Figure 17: Private Mode Trips Produced for the Year 2025

The projection for 2025 follows the same trend as the 2015 map, with the Horto macro-area (MZ 10) beginning to stand out.

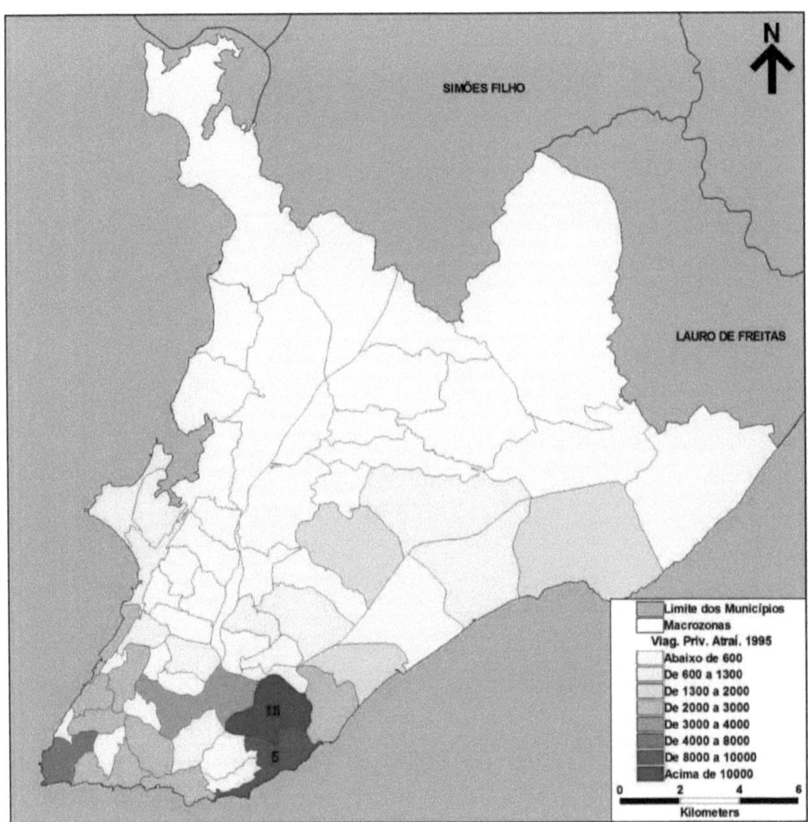

Figure 18: Attracted Private Mode Trips for 1995

As for trips Attracted by Private Mode in 1995, the biggest highlights are Iguatemi (MZ 15) and Pituba (MZ 5), regions that concentrate jobs and a higher-income population, as shown in Figure 7 and Figure 5.

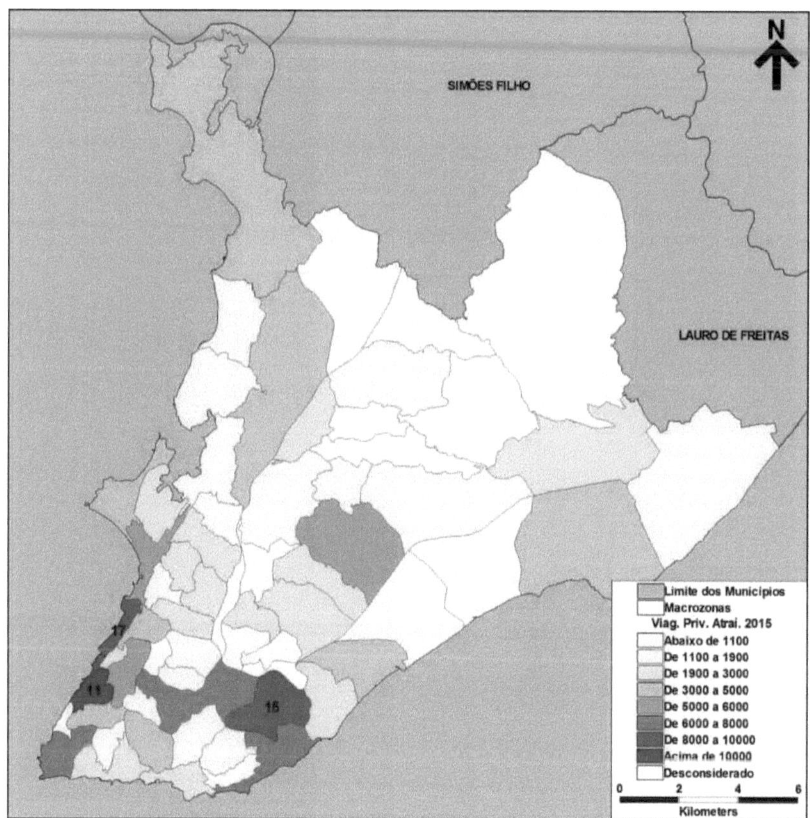

Figure 19: Attracted Private Mode Trips for 2015

For 2015, there is still a trend towards attraction in the Iguatemi region (MZ 15), but with greater attraction for the central region of Campo Grande (11) and Comércio (MZ 17).

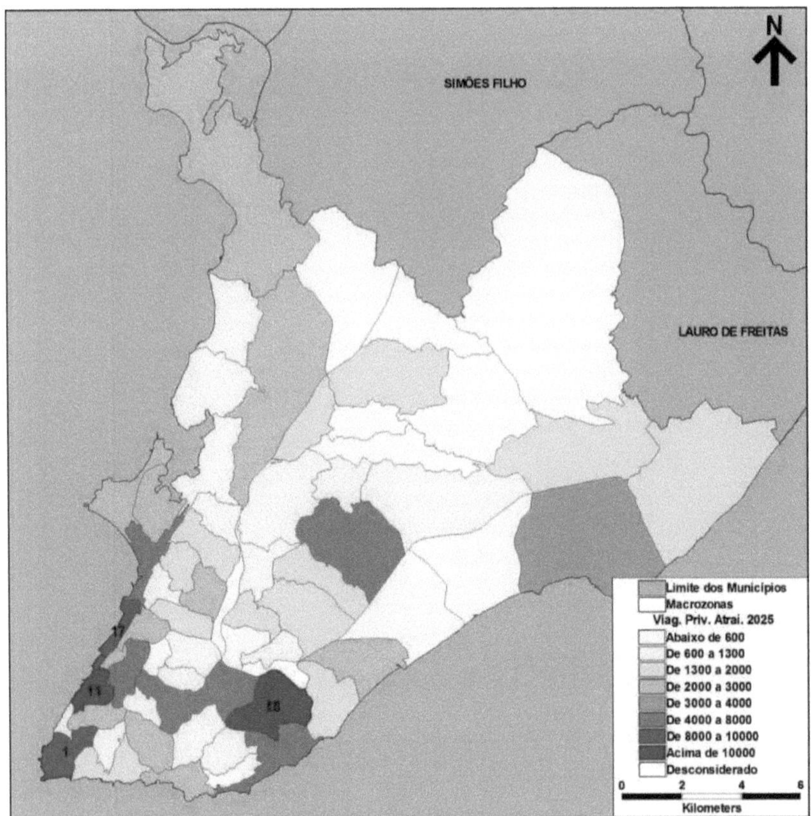

Figure 20: Attracted Private Mode Trips for 2025

The projection for 2025 follows the same trend as the 2015 map, with Iguatemi (MZ 15) becoming more prominent and Barra (MZ 1) starting to stand out.

4.   3Future guidelines

The most appropriate journey distribution model was chosen by comparing histograms relating distance and number of journeys, as shown in the following figures.

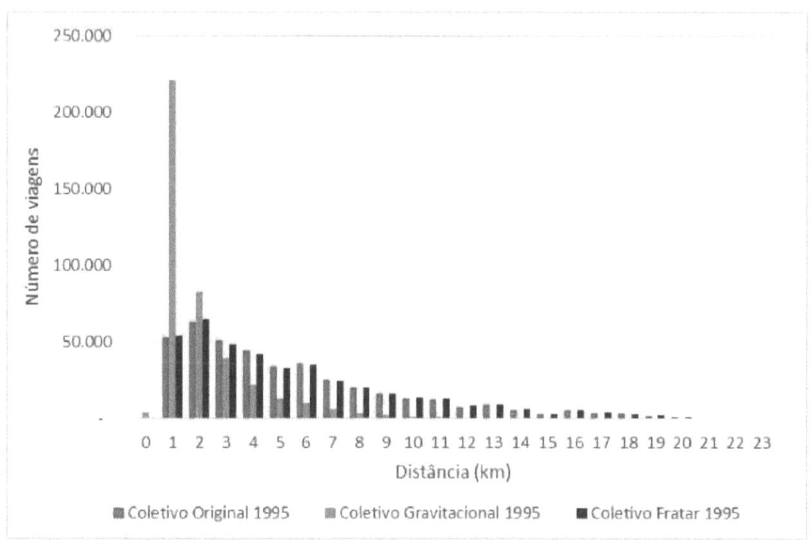

Figura 21:   Histogram of number of journeys by distance (km) for the Collective mode comparing the number of journeys in the base year 1995 with the number of journeys estimated by the Fratar and Gravitational models also for 1995

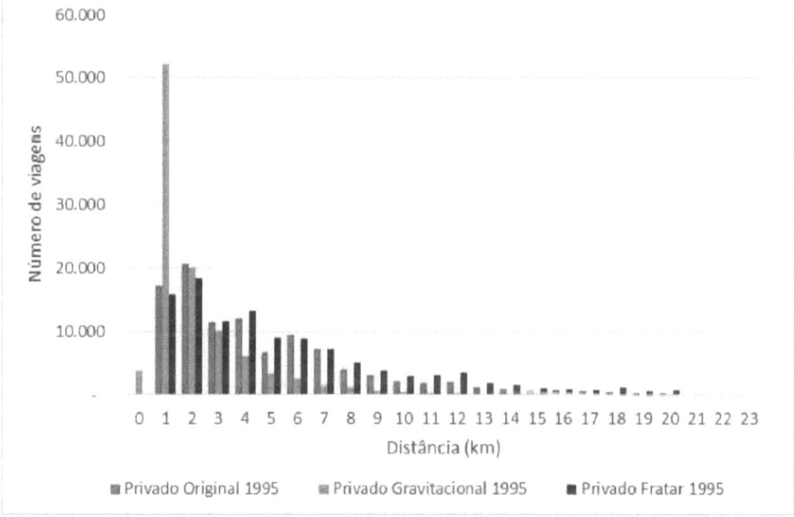

Figure 22: Histogram of number of journeys by distance (km) for the Private motorised mode comparing the number of journeys in the base year 1995 with the number of journeys estimated by the Fratar and Gravity models, also for 1995.

It can be seen that the distribution of journeys versus distance for the Fratar model is similar to the distribution for the base year. In the Gravitational model, macro-areas with small distances between each other have a much higher number of journeys, but this is not the predominant decision factor, so the model did not represent reality well.

The Fratar model was therefore chosen for the journey distribution stage.

4.   4Desire Line Maps

The resulting desire lines for the base year and the project years are shown below.

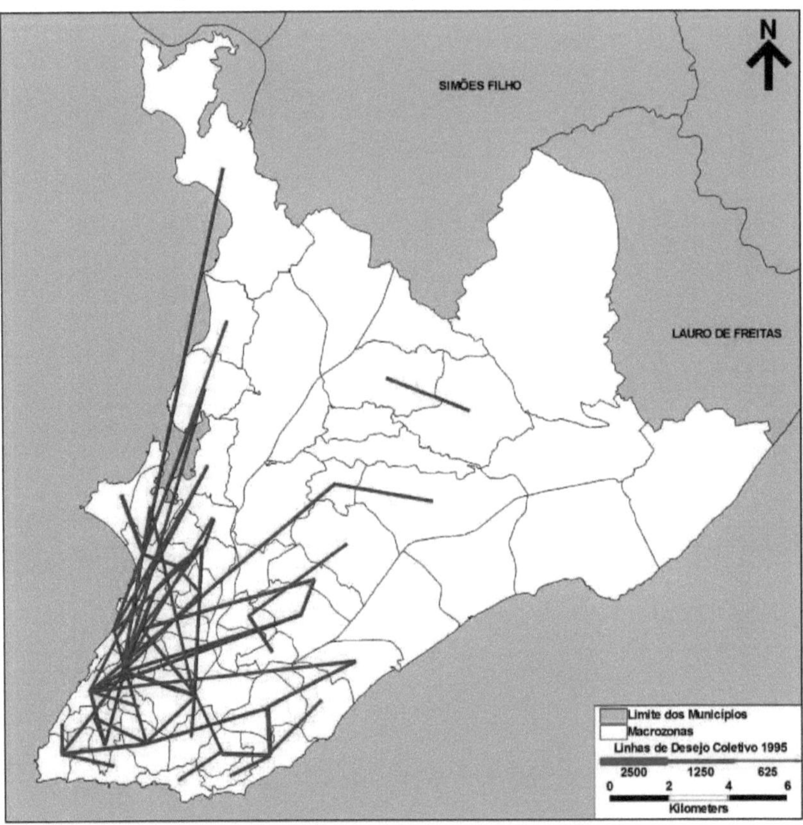

Figure 23: Collective Mode Desire Lines for 1995

For the Coletivo mode in the base year of 1995, there was a predominance of desire lines in the Centro and Liberdade regions. At the time, the macro-areas that concentrated the main flows of attraction and production of trips are listed below:

Table 7: Macrozones with high production and attraction of trips by Collective mode in 1995

Producers			Attractors		
Order	MZ	Name	Order	MZ	Name
1°	14	Brotas	1°	11	Campo Grande
2°	38	Massaranduba	2°	12	Nazareth
3°	33	Retiro Farm	3°	17	Trade
4°	51	Coutos	4°	15	Iguatemi
5°	67	Tancredo Neves	5°	5	Pituba

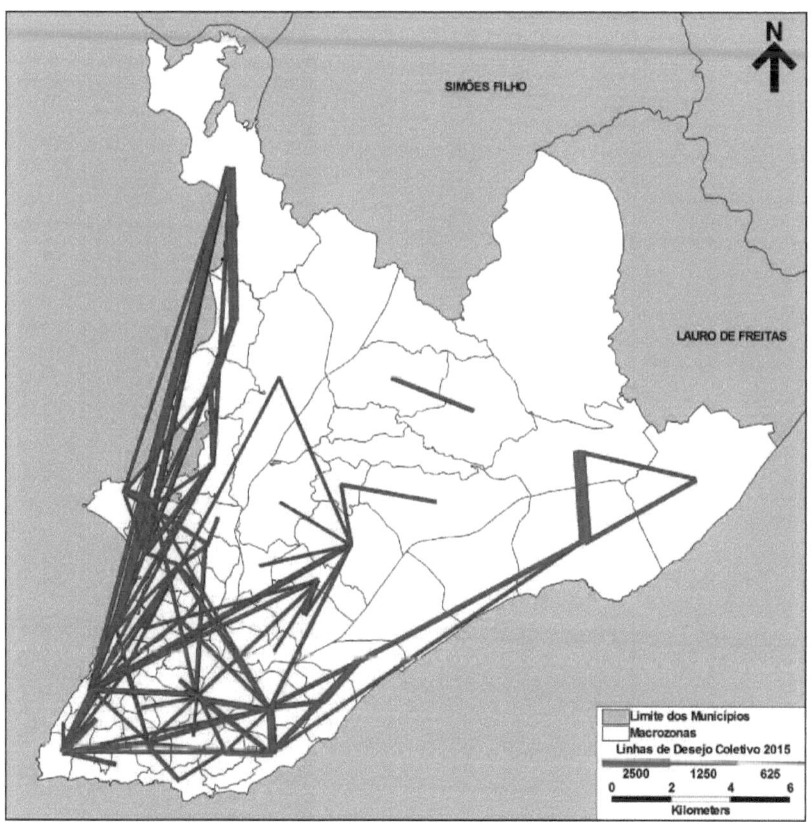

Figure 24: Collective Mode Desire Lines for 2015

In 2015, some other lines of desire began to emerge beyond the Suburb-Centre axis. The relations between Iguatemi and Brotas with other macro-areas such as Pituba, Barra, Liberdade and Orla Norte stand out. In this figure you can see the consolidation of the new centre, close to the Iguatemi Shopping Centre and the Bus Station, as well as the Brotas region, where BR-324 and Paralela Avenue converge.For 2015, the macro-areas that stand out most in terms of number of trips are:

Table 8: Macrozones with high production and attraction of trips by Collective mode in 2015

Producers			Attractors		
Order	MZ	Name	Order	MZ	Name
1°	38	Massaranduba	1°	11	Campo Grande
2°	51	Coutos	2°	15	Iguatemi
3°	33	Retiro Farm	3°	17	Trade
4°	14	Brotas	4°	1	Barra
5°	32	Freedom	5°	5	Pituba

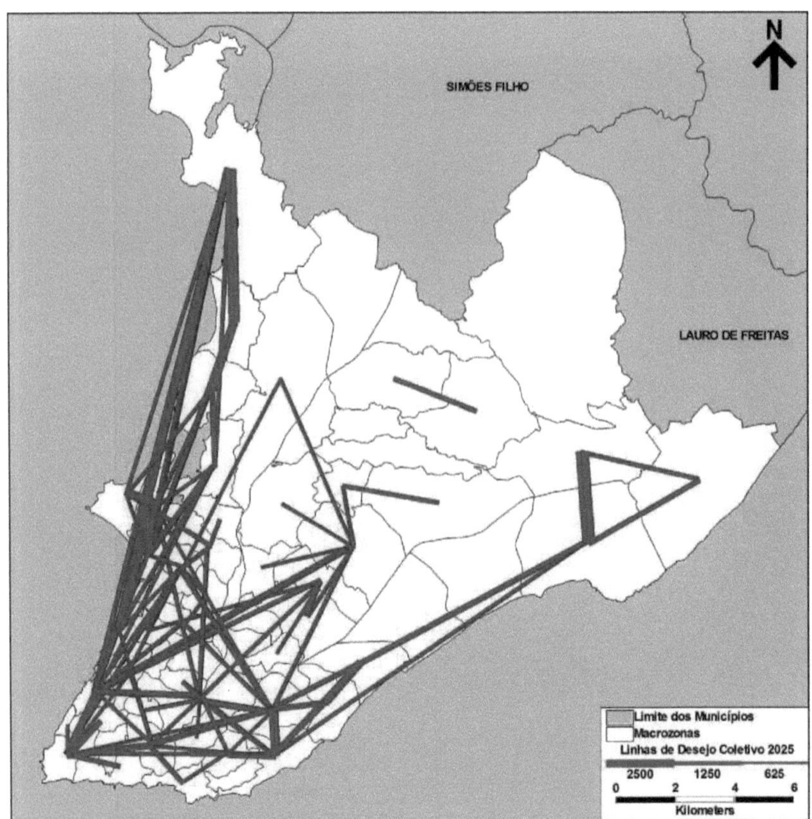

Figure 25: Collective Mode Desire Lines for the year 2025

For 2025, the same trends can be seen as in 2015, but the lines of desire are heavier.

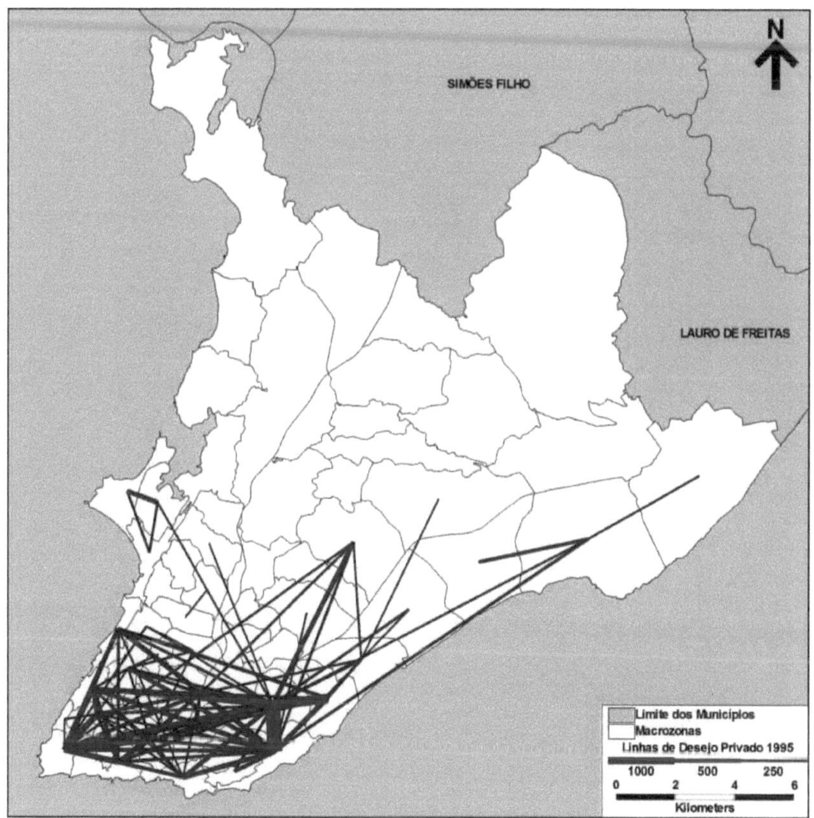

Figure 26: Private Mode Desire Lines for 1995

For the Private mode in the base year of 1995, almost all the flows were between the regions of Iguatemi, Pituba, Barra, Campo Grande and Comércio, which can be clearly seen in the following table.

Table 9: Macrozones with high production and attraction of trips by Private Motorised mode in 1995

Producers			Attractors		
Order	MZ	Name	Order	MZ	Name
1°	15	Iguatemi	1°	15	Iguatemi
2°	5	Pituba	2°	5	Pituba
3°	1	Barra	3°	1	Barra
4°	14	Brotas	4°	14	Brotas
5°	11	Campo Grande	5°	11	Campo Grande

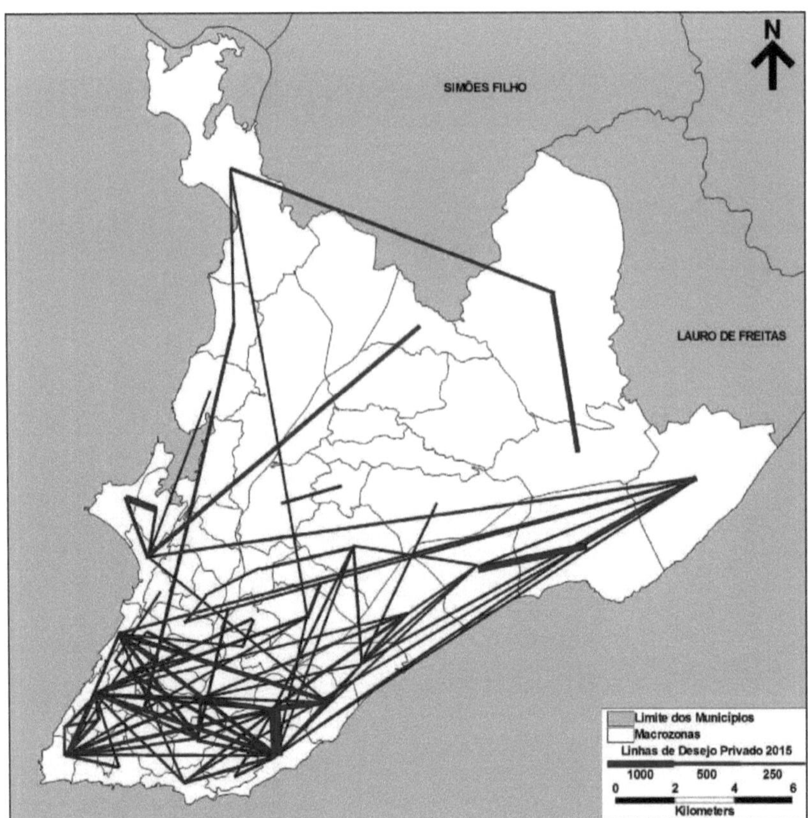

Figure 27: Private Mode Desire Lines for 2015

For the Private mode in 2015, there were large flows between Iguatemi and Pituba, while regions such as Comércio, Campo Grande, Barra and Stela Maris also featured prominently. In this way, we can clearly see the axis of expansion of the Orla Norte and its characteristic of prioritising the private mode due to the income of the population living there. The macro-areas that stand out in terms of the number of journeys are shown below:

Table 10: Macrozones with high production and attraction of trips by Private mode in 2015

Producers			Attractors		
Order	MZ	Name	Order	MZ	Name
1°	5	Pituba	1°	11	Campo Grande
2°	15	Iguatemi	2°	17	Trade
3°	47	Stela Maris	3°	15	Iguatemi
4°	10	Garden	4°	1	Barra
5°	16	C. Conventions	5°	14	Brotas

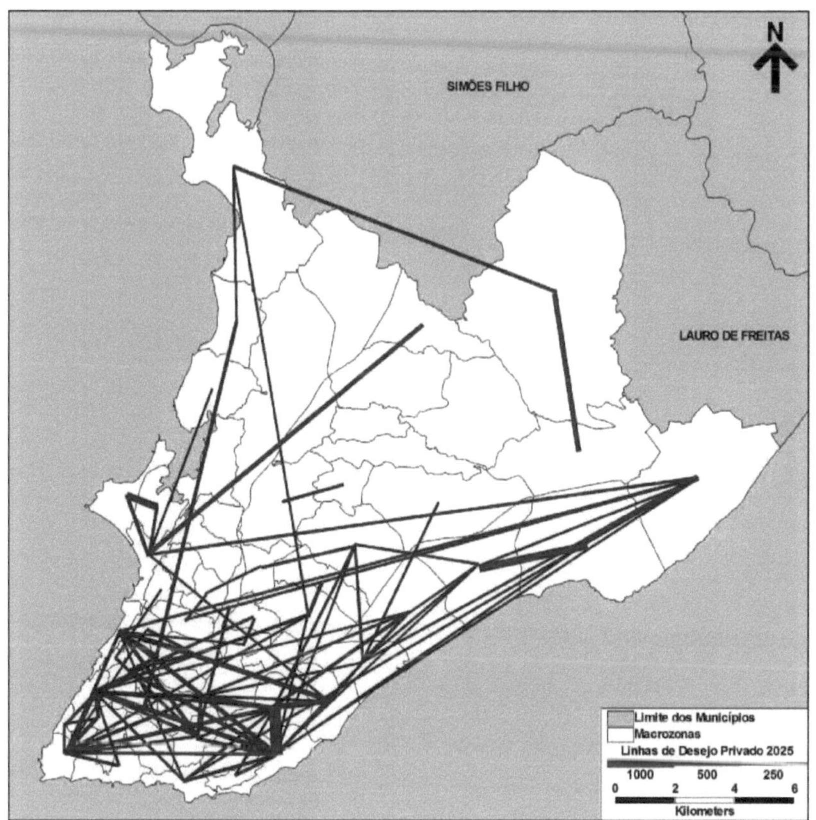

Figure 28: Private Mode Desire Lines for 2025

For 2025, the same trends can be seen as in 2015, but the lines of desire are heavier.

4.5Problems with   the current network

The current network is shown in Figure 29.

Figure 29: Salvador's Current Transport Network (Source: CCR Metrô Salvador and Trem de Subúrbio de Salvador)

The section of the metro from Pirajá to Águas Claras/Cajazeiras is under construction, while the entire section of line 2, from Bonocô to the airport, is still in the design phase.

Comparing the current network with the thematic map of total journeys produced by collective mode from 1995, Figure 9, it can be seen that it serves regions of high production. The metro serves Brotas (MZ 14) and Fazenda Grande Retiro (MZ 33), while the suburban train serves Liberdade (MZ 32), Fazenda Grande Retiro (MZ 33), Massaranduba (MZ 38) and Coutos (MZ 51).

It should be noted that although the suburban train serves several macro-areas, it has a lower performance than the metro.

In the map of journeys produced by public transport in 2015, Figure 10, the Malvinas region (MZ 36) on the northern fringe and some macro-areas in Miolo Norte, Subúrbio Sul and Pirajá are beginning to gain prominence. The same trend is observed for 2025.

Despite the plans for Line 2 of the metro, linking Bonocô to the airport, this has not yet been implemented. As can be seen from the maps, at the present time, in 2015, there is already a demand on the northern fringe that is not adequately met.

With regard to the trips attracted, the map of trips attracted by public transport in 1995, Figure 12, shows that the Campo Grande region (MZ 11) stands out, as well as some other attraction macro-areas with less impact in the Centre, Brotas, Pituba, Península and Federação regions.

The current line serves the Campo Grande region (MZ 11), but leaves something to be desired in other macro-areas.

In the map of journeys attracted by public transport in 2015, Figure 13, it can be seen that some of the macro-areas mentioned above are consolidating, with the highlights being Campo Grande (MZ 11), Iguatemi (MZ 15) and Comércio (MZ 17).

According to the projected network, Iguatemi and Comércio are reasonably well served, as they are close to the network. Considering that the stretch of Line 2 has not yet been built, the Iguatemi macro-area still lacks adequate transport.

Analysing the lines of desire of the 1995 Collective mode, Figure 23, it can be seen that the greatest flows are directed between Campo Grande and the sub-regions of Subúrbio, Península, Miolo and Federação.

In this way, the current network does not adequately cater for flows, since there is no direct way of travelling from Subúrbio to Campo Grande, you must necessarily use the train and bus or train, bus and metro. The Miolo and Federação regions are also not adequately served, as Line 2 has not been built and there is no connection to Federação.

As for the lines of desire of the Collective mode in 2015, Figure 24, the greatest flows are still observed in the connection between the Suburbs and the Centre and new lines of desire stand out in the Orla expansion axis and for the connections between Brotas, the Centre, Pituba and Federação.

Thus, by evaluating public transport journeys, it can be concluded that the current network serves important regions such as Campo Grande, Península and the Subúrbio Ferroviário, and intends to serve other prominent regions such as Iguatemi and Orla Norte. However, there isn't the necessary integration to connect the Centre and the Suburbs, and regions such as Federação and Pituba don't have adequate transport.

As for the trips produced by private mode in 1995, Figure 15 shows that the Iguatemi (MZ 15) and Pituba (MZ 5) regions stand out. The former will have a reasonable service once Line 2 of the metro is built, but the latter has been ignored by current planning.

In the map of journeys produced by private mode in 2015, Figure 16, it can be seen that, in addition to the regions mentioned for 1995, Stela Maris (MZ 47) stands out, which will have a considerable service once Line 2 of the metro is built.

For 2025, the behaviour is very similar, with Horto (MZ 10) starting to stand out. Thus, the sub-regions of Pituba, Brotas and Orla Norte concentrate trips produced by private mode and none of these regions are currently adequately served.

When it comes to trips attracted by private mode in 1995, Figure 18 again highlights Pituba (MZ 5) and Iguatemi (MZ 15), with the same problems mentioned above.

In terms of trips attracted by private mode in 2015, Figure 19, Campo Grande (MZ 11), Comércio (17) and Iguatemi (MZ 15) stand out. The problem with Line 2 is that although it was designed to meet the needs of Iguatemi, it has not yet been built.

Analysing the desire lines by private mode in 1995, Figure 26, shows a predominance in the regions of Iguatemi, Pituba, Barra, Campo Grande and Comércio. These areas are not adequately covered by the current network.

The desire lines by private mode in 2015, Figure 27, reflect an expansion of journeys by private transport, previously concentrated only on the axes described in 1995.

It can be seen that the current line does not serve the regions where there is the greatest demand for private transport. Line 2 is expected to serve the Orla Norte region, but even so, the Federação and Pituba sub-regions still lack connections.

Thus, private transport users will not migrate to public transport, initially because the current network does not meet their needs. Obviously, there are other factors that drive this switch, such as the quality of the service, but this initial analysis shows that the first requirement for migration to take place is not being met.

Therefore, it can be concluded that the current and projected network is adequate in parts. It provides good coverage of the city's expansion axes according to its needs, such as Orla Norte and Subúrbio Ferroviário, but ignores important connecting stretches so that users can really get where they want to go, as in the case of the link between Subúrbio and Centro.

The current network covers public transport users. In order to include private transport users, some changes would have to be made.

Its design follows valley avenues and is elevated for most of its route, so it was built in the "easiest" places and in the least costly way possible.

Also noteworthy is the delay in the metro works. The Line 2 project is very suitable, as it connects the airport and the northern waterfront to the centre, but if it is not implemented, no changes will be made to municipal transport.

Finally, although the metropolitan region is not analysed in this project, it is worth noting that the city of Lauro de Freitas is conurbed with Salvador and is an important hub that could be connected to the network.

4.   6Proposed network

As discussed in the previous section, the current network serves the municipality well, but has some flaws. It was therefore decided to adopt the current network as the basis for the proposed network, adding the expansion axes that were deemed necessary.

It was also decided to use the current network for construction reasons, since it is already located in the easiest places to install, the valley bottom avenues. In this way, even if other routes were proposed, the relief would direct the lines to very similar routes to those already adopted.

Thus, four axes of expansion were drawn up to serve regions that the current network does not cover:

-Connection   between Barra and Lapa
-Connection   between Pituba and Iguatemi
-Connection   between the railway line and the metro in the Fazenda Grande area
-Connection   between the airport and Lauro de Freitas, a neighbouring municipality
        conurbed with Salvador.

The proposals can be seen in Figure 30 below.

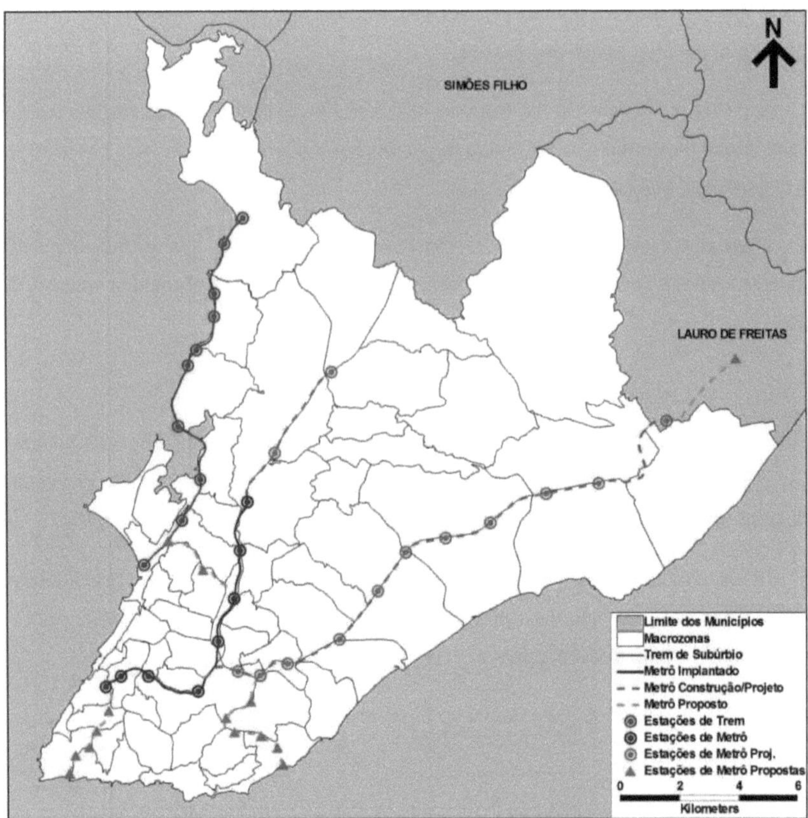

Figure 30: Proposed transport network

The stretches from Barra to Lapa (Shopping Barra - Lapa) and from Pituba to Iguatemi (Praça Nossa Senhora da Luz - Rodoviária/Iguatemi) are shown in detail in Figure 32 below.

The Shopping Barra - Lapa stretch follows the route of Av. Centenário and Av. Vale do Tororó, so it can be elevated or underground on the Av. Centenário stretch and preferably underground on the Av. Vale do Tororó stretch.The Praça Nossa Senhora da Luz - Rodoviária/Iguatemi stretch follows Rua Pernambuco and Av. Antônio Carlos Magalhães, an important bus corridor in the region. Ideally, the stretch of Rua Pernambuco should be underground, while the stretch of Av. Antônio Carlos Magalhães can be elevated or underground, as the avenue is wide and appears to be able to accommodate both technologies.

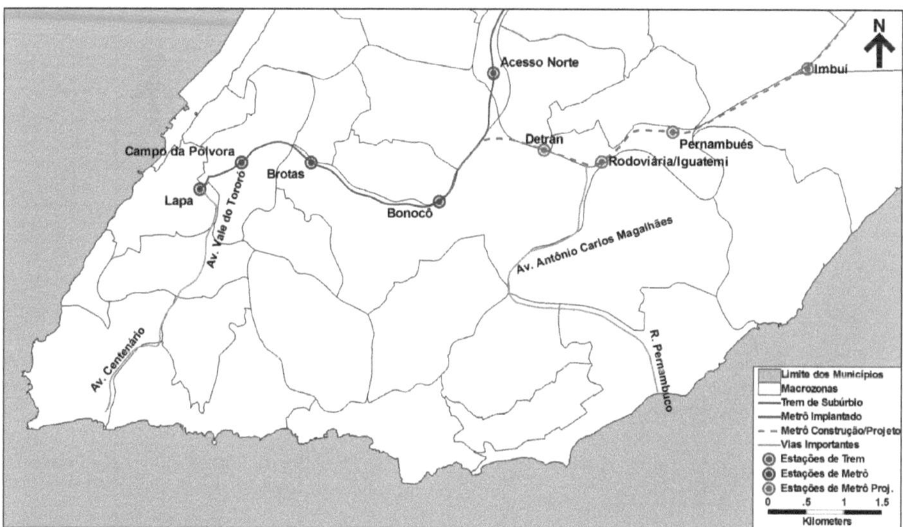

Figure 31: Detail of the South Region's current transport network and main routes commented on

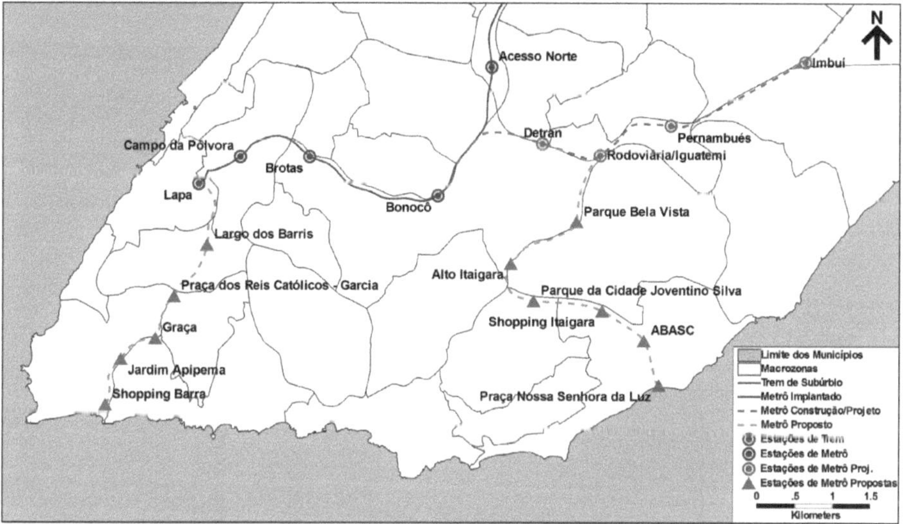

Figure 32: Detail of the southern region of the proposed transport network

The connecting section between the train and the metro is called Baixada do Fiscal - Retiro. The Baixada do Fiscal station is also a proposal for the existing railway line. The section would follow the route of General San Martin Avenue and should preferably be underground

Due to the great difference in slope observed in the Baixada do Fiscal region, the ideal would be to build a station on different levels, one at ground level for the train and a higher, elevated one for the metro.

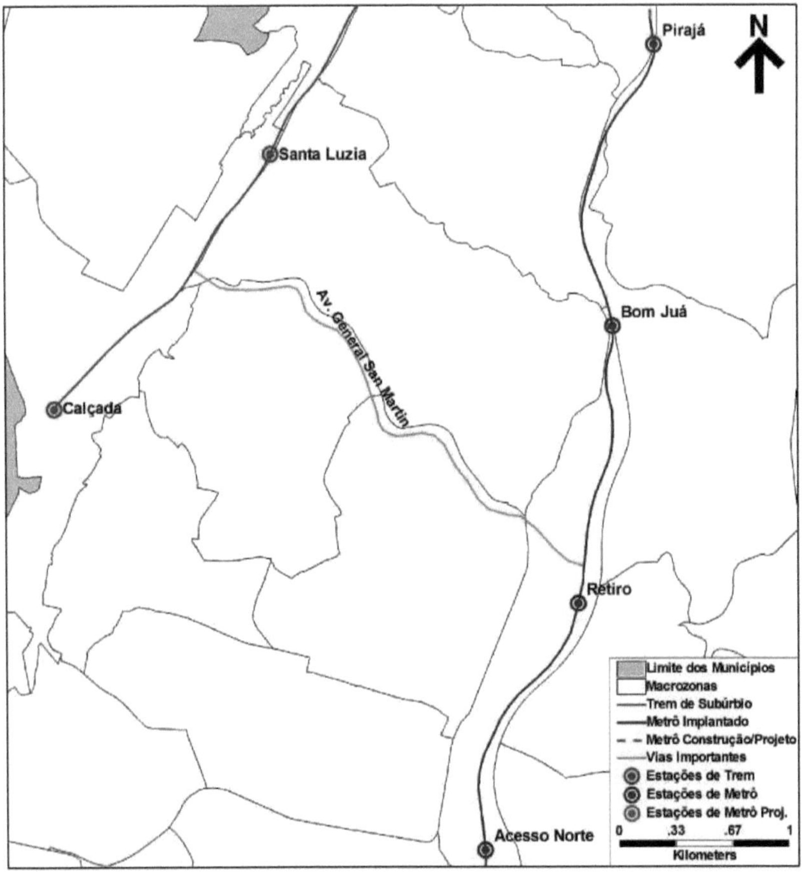

Figure 33: Detail of the connection between the Suburban Train and the current transport network and the main commented route

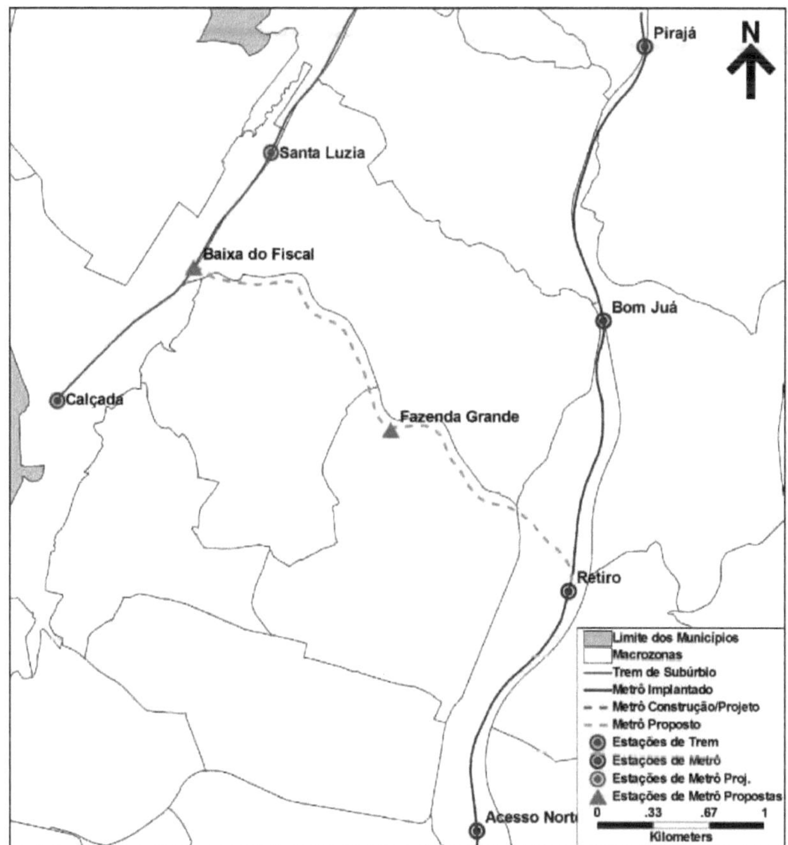

Figure 34: Detail of the connection between the Suburban Train and the proposed transport network

As previously mentioned, the ideal network would have a branch of the metro arriving in Lauro de Freitas city centre, passing through Estrada do Coco. The section could be elevated, like the one planned for Luís Vianna Avenue.

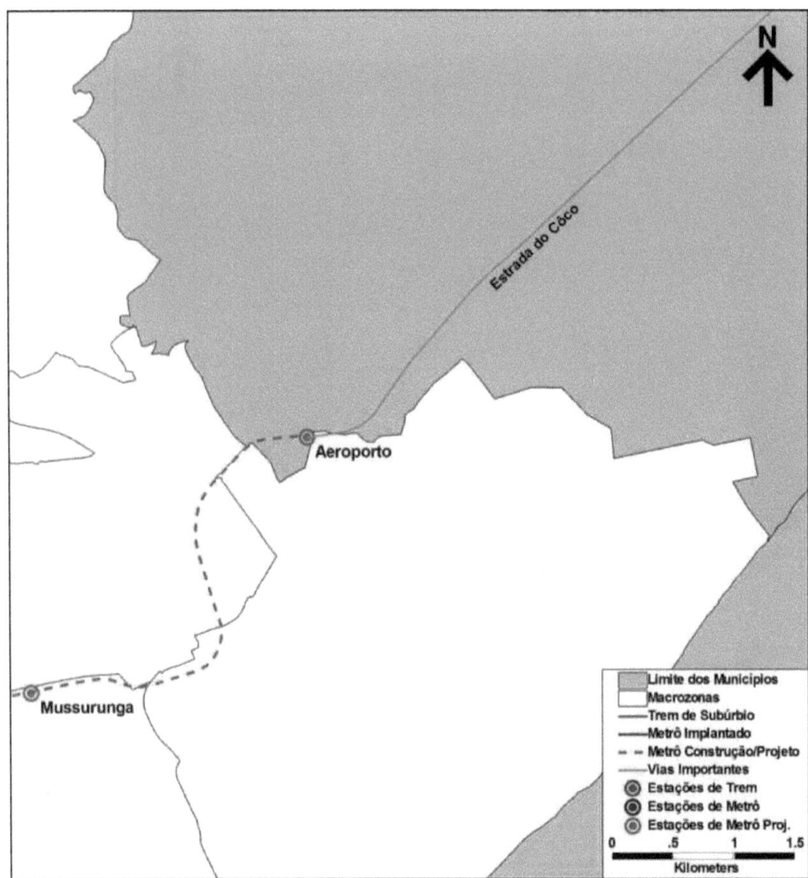

Figure 35: Detail of the Lauro de Freitas section of the current transport network and main road commented on

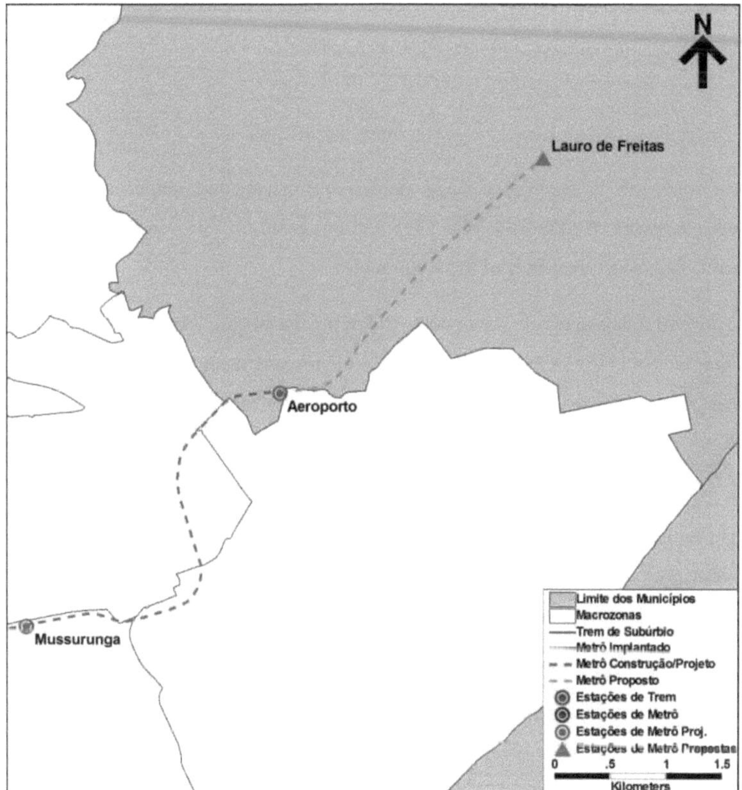

Figure 36: Detail of the Lauro de Freitas section of the proposed transport network

Finalising the proposal, the ideal would be for the train system to be reformed and modernised into a metro system, or at least to improve its performance, especially in terms of speed and passenger capacity.

In this way, proposals could be made for lines that connect the suburbs to the central region quickly and efficiently, with the user only having to make one transfer at Baixada do Fiscal due to the terrain.

CHAPTER 5

Conclusions

The aim of this study was to analyse passenger demand for private and public transport for the Salvador - BA metro network, for the base year 1995 and projected for 2015 and 2025, and to propose a metro network that meets the needs of the study region.

Analysing the current metro network, it can be seen that it is installed along the city's axes of expansion, on roads that have already been consolidated, and its implementation has been favoured by the relief and configuration of the roads. However, it fails to serve important regions, such as Barra and Pituba, and does not include important connections, such as from the Railway Suburb to the Centre.

As well as meeting the demand for public transport, the proposals on offer aim to reach users of private motorised transport, so that by offering quality public transport, such as the metro, these users migrate to the new system.

Demand analysis is the initial stage in analysing the feasibility of a new metro line. Technical, operational, geotechnical and environmental studies, as well as a financial assessment, among others, are still needed before the actual implementation can begin.

In this way, this work is the start of the feasibility analysis, providing initial material for further studies, which, if carried out, should make up a feasibility project for Salvador's metro network.

Bibliographical references

BARBETTA, Pedro A.; Statistics Applied to the Social Sciences. Florianópolis: Ed. UFSC, 2012.

BRUTON, Michael J.; Introduction to Transport Planning. São Paulo: University of São Paulo Press, 1979.

CAMPOS, Vânia Barcellos Gouvêa; Transport planning: Concepts and analysis models. s.l. : Handout in digital format, 2007.

CARVALHO, Inaia Maria Moreira de; PEREIRA, Gilberto Corso and RIBEIRO, Luiz Cesar de Queiroz; Salvador: transformations in the urban order. Rio de Janeiro: Letra Capital, Observatório das Metrópoles, 2014.

GORDILHO-SOUZA, Ângela; Limites do habitar: segregação e exclusão na configuração urbana contemporânea de Salvador e perspectivas no final do século XX. Salvador : EDUFBa, 2000.

KAWAMOTO, Eiji; Transport Systems Analysis. São Carlos : Graphic Service EESC - USP, 2010.

MENDES, Victor M. O.; A problemática do desenvolvimento em Salvador: Análise dos planos e práticas da segunda metade do século XX (1950-2000). Thesis (Doctorate in Urban and Regional Planning) Federal University of Rio de Janeiro, Rio de Janeiro, 2006.

ORTÚZAR, Juan de D.; WILLUMSEN, Luis G. Modelling transport. Chichester: John Wiley & Sons Ltd, 2011, 4 ed.

PLAZA, Conrado V.; Intrazonal travel distance: estimation approach and application to a case study. Thesis (Master's in Transport Engineering) - São Carlos School of Engineering, University of São Paulo, São Carlos, 2014)

ROCHA, Samille S.; Analysis of Urban Travel Generation by Public Transport through Geostatistical Techniques. Thesis (Master's in Urban Environmental Engineering) - Polytechnic School, Federal University of Bahia, Salvador, 2014.

Natália Fenner Pena

Phase Angle, Nutritional Status and Clinical Outcomes in Oncology

Printed by Books on Demand GmbH, Norderstedt / Germany